I, TOO, MUST CRY

The Story of
Major John David Gilmore
Vietnam Veteran – Missing in
Action

H. Douglas Bankhead

Content Warning:

This novel expresses the violence of war, sexual and religious content.

Published in the United States of America by Pegg Legg Publishing

Second Edition
ISBN: 979-8-9915160-0-6
EBook ISBN: 979-8-9915160-1-3

...To the thousands of UNITED STATES SOLDIERS who gave their life so that the Vietnamese people could live under a free Democratic form of government.

To the hundreds of AMERICAN SOLDIERS who chose, for various reasons, to remain in Vietnam.

A special thanks to Mr. Mark H. Wilson, President of Orthotic & Prosthetic Design in St. Louis, MO, and his staff who so kindly dedicates their God given skills as prosthetic professionals to keep disabled veterans, like myself, to be independent to pursue their dreams and aspirations.

... And to my wife, Tina, my professional editor who made the publication of "I, Too, Must Cry" possible.

... And my son, Douglas, and his wife, Amber – Thanks for the encouragement ... "Yes! You can Dad!"

Contents

Prologue

Vietnam: The War Continues!

The Vietnam War distracted, delayed, and destroyed the lives and families of a generation of American young men and women. Thousands never recovered. Suicide became a permanent solution to a treatable, and temporary diagnosis of post-traumatic stress disorder, referred to as PTSD. Hundreds became homeless and addicted to drugs and alcohol, which was a slow death of intentional suicide.

And yet, a few who became mothers and fathers of Amerasian children decided to remain in Vietnam to raise their families.

Major John David Gilmore was one of those few who chose to be counted as missing in action (MIA) so that he could protect his Vietnamese wife, and help raise their three children and her two teen-aged brothers.

Introduction

J ohn David Gilmore was born to the parents of Alonzo and Rose Gilmore. He was the sixth of seven children born to Mr. and Mrs. Gilmore. Rose died giving birth to her seventh child, a girl named Rachel.

John David Gilmore died July 26, 2021, in Cu Chi Viet Nam. He died on his birthday, and was laid to rest on the property which he shared with his Vietnamese wife (Hao Hong Nguyen-Gilmore) John David Gilmore was eighty. Gilmore held the rank of major. He was a jet pilot and was a trained military intelligence officer, assigned to give support to the United States Air Force.

Major Gilmore was reported missing in action (MIA) on 31, January 1968. He was not MIA. He chose to stay in Vietnam. Major Gilmore had a Vietnamese wife and three children, in Cu Chi, a small village south of Saigon, Vietnam.

From the month of January 1970, until his death, Major Gilmore and I stayed in communication with each other through regular United States Postal Service mail, and an occasional telephone conversation. I served as Major Gilmore's door-gunner, and chief mechanic during his service in Vietnam.

During the numerous conversations that we shared through the years and decades, Major Gilmore requested of me to publish his life story, with the sacred promise that his father's life story would precede his in the publication. I have kept that promise!

As you are reading this true story, with only the names being changed to protect the privacy of those who are still alive, you will be moved and probably be changed emotionally concerning the Vietnam War, and the American soldiers who chose not to return to their life and homes to the United States of America!

Chapter One

Missing In Action? No.

The war between the United States of America (USA) and the North Vietnamese Army (NRA) ended on April 30, 1975. The US Congress considers the Vietnam Era to be the period beginning on February 28, 1961, and ending on May 9, 1975. During that period of fighting, approximately 58,000 US Military soldiers were killed in action (KIA), and an additional 304,000 were wounded. *What a waste!*

After the ending of the Vietnam War, human collateral and damages continued to persist. US personnel, both military and civilian, parented an estimated 10,000 to 20,000 children. This population of American children are identified as Amerasian. Their future continues to be bleak, as they are discriminated against when it comes to education, employment, housing, and access to medical care. Major Gilmore accepted his responsibilities when he chose to father a child. He married the Vietnamese woman who gave birth to his daughter, and two additional children, fraternal twins.

By accepting these responsibilities, Major Gilmore gave up his opportunities of how he would be rewarded as a hero and decorated soldier. Political and business opportunities surely awaited him.

Why would a decorated military pilot and intelligence officer choose to remain in Vietnam to be counted as missing in action?

Major John David Gilmore completed his college education with honors, he graduated number one in his flight class of seventeen, and the National Aeronautical Space Administration (NASA) had accepted him to attend astronaut training upon completing his tour of duty as a fighter jet pilot and intelligence officer. Major Gilmore was being recruited by both the Republican and Democratic parties to join their affiliation to run for a high political office, following his military service.

The challenges and hardships of his father, Alonzo, and his mother, Rose, suggest that their life experiences, as they shared with their dear children, strongly influenced his decision to remain in Vietnam to support the mother of his children and raise their family.

While there are still (as of this writing) 1,244 American personnel MIA from the Vietnam War. Major John David Gilmore was never MIA. He chose to stay in Vietnam to provide, protect, and raise his family. He did what any honorable husband and father would do.

Chapter Two

The Early Years

I t was a cool and frosty morning in late October. The year was 1947. John David Gilmore was six years old. John David stood at a window of his mother's bedroom watching his brothers and sisters walk the two miles to their one-room school, Rush Ridge Elementary School. The school was segregated and had an enrollment of approximately thirty students. There was one teacher, who taught grades one through eight. The children who attended Rush Ridge Elementary School were all residents of the Village of Mule Ridge, which was primarily made up of negro (who are now identified as "black American") families, who provided labor for the production of cotton.

There were other villages that made up the southeast Missouri Cotton Belt. Among them were Bird's Point, near the Illinois border, Brewer's Lake, Wilson City, and Alfalfa Center. Each of these villages had their separate school district.

The children were happy to be back in school. They had been out of school for what was known as the cotton-picking season, which lasted from the middle of August to the last Monday of October. There was also a cotton copping season, a process of removing weeds and grass from the cotton plants. This season began around the last of April and ended in the middle of August. With the closing for the traditional summer break and the Thanksgiving and Christmas holidays, segregated schools in the Cotton Belt of southeast Missouri were in session for approximately five months out of a year.

John David was being home schooled by his older sisters and brothers. Martha, the second oldest daughter, was John David's primary home-school teacher. "She was a strict and no-nonsense teacher," John said of Martha.

As John David watched his siblings walk to school, they would meet Rose, their mother, as she was returning to their home from her job at the Smithfield mansion, which was approximately halfway between the Gilmores' home and the one-room schoolhouse. Although Rose owned an automobile, and was a licensed driver, she chose to walk to and from work, when weather permitted, to maintain her physical fitness. The walk would also allow her to meet and greet her beloved children on their walk to school.

Standing at the window, John David's breath would fog the window pane where he would take his right index finger and write "LOVE MOM!" Martha had taught John David how to write that endearing greeting.

As Rose met and greeted her beloved children, she would present each with a candy bar, which had been given to Rose by Mrs. Adaline for this moment of mother and children bonding. Before parting from her children and reminding them to study their lessons and be obedient to their teacher, Rose reminded them that it was the season for gathering pecans, hickory nuts, and black walnuts, and for them to return home from school as fast as they could, so that they could have a lot of day light for gathering the nuts.

Upon arriving at home, Rose would greet and hug John before presenting him with his candy bar. Rose believed that John would be the last time for her to give birth to another baby, as the period between her other children was eighteen months to two years. She was wrong. Rose was pregnant with her seventh child!

Chapter Three

Alonzo and Rose Reminisce

R ose so looked forward to the season of gathering the various types of nuts and wild fruits: pecans, hickory nuts, black walnuts, persimmons, and wild grapes. The nuts and wild fruit gathering season signaled the beginning of the holiday festivities and celebrations, which began with the Wyatt annual parade and carnival, Thanksgiving, and finally Christmas.

Rose would use the bounty of nuts and wild fruits to bake pecan pies, persimmon bread, canning of wild grape jellies, and jams. Rose had a special processing of black walnuts and hickory nuts to use in her homemade ice cream recipes

After preparing a picnic style meal for the nut gathering outing with her beloved children, Rose turned her attention to preparing lunch for her husband who was out gathering fuel wood for the winter months. As the home of Alonzo and Rose was fueled by propane gas for both heating, cooking, washing of clothes, as well as heating water for the bathroom, Alonzo harvested the wood to use when heating water for the butchering and curing of meat. In addition to these requirements, much of the wood was harvested for sale, as it added to their cash flow.

Upon arriving home from his chores of gathering fuel wood, Alonzo and Rose greeted, hugged, embraced, and kissed passionately. Alonzo and Rose loved each other completely and dearly.

As the committed husband and wife sat at the dinner table enjoying the meal of mashed potatoes, green beans, and corn bread, Rose confirmed to her husband that he would soon be the father of their seventh child. "I have missed my menstrual cycle for the second month in a row," Rose happily shared with her husband and father-to-be.

Responding with a loud and prideful shout of joy, Alonzo embraced his dear wife and mother-to-be with happy outbursts. "Great! Wonderful. How God is blessing our family!"

While discussing their respective days of activities and having fun chit chat over the meal that Rose lovingly and carefully prepared, Rose had an irreversible yearning to suggest to Alonzo that it was time for them to share with their children the stories of their life's journey and experiences that their mom and dad had overcome. Alonzo nodded his head in the affirmative.

"Alonzo, honey, we can't tell our children, in detail, about our courtship, petting, and the day and night of our honeymoon. You can share that with the boys, when they are older." Rose said to her smiling and proudful husband, as they hitched Blue to the family wagon, which they used for the occasions of gathering nuts, and family picnics.

Alonzo first laid his eyes on Rose at the memorial service for Mr. Hunter Smithfield, which was held at the negro Baptist Church of Mule Ridge. It was the month of October 1927. Alonzo was approximately twenty-three years old, and Rose was approaching the age of fourteen, but had the physical attributes and maturity of someone older. She remained in the mind's eye of Alonzo.

Rose was born on February 3, 1914. Charlotte Rose Broussard was Rose's mother, who was fifteen years old when Rose was born. Rose never knew her father. However, Mabel Broussard, Rose's grandmother, believed and pondered in her heart that "Little Hunter" was the father of Rose. The two who grew up together always played and frolicked together, even into their teen aged years.

Andrew and Mabel Broussard, Rose's grandparents, raised Charlotte Rose in an unsettled Creole hamlet, approximately two miles south of New Orleans, Louisiana.

Andrew and Mabel scratched out a lively hood by working for the Smithfield family. Andrew worked the cotton fields, and Mabel labored as the housekeeper for Eugene Smithfield, the father of Hunter Smithfield.

Eugene Smithfield purchased the Mule Ridge Plantation in the year of 1900. Hunter and his wife, Adaline, moved to southeast Missouri after the purchase of the Mule Ridge Plantation. Hunter and Adaline were the legal owners and managers of the Mule Ridge Plantation.

Andrew and Mabel maintained their employment with Mr. Smithfield, and—with their daughter Charlotte Rose—they moved to southeast Missouri with Mr. Smithfield and his wife when they purchased the Mule Ridge Plantation.

Like Andrew and Mable, who had one child, Charlotte Rose, Mr. Hunter and Mrs. Adaline had one child, a son by the name of Hunter Smithfield Jr., who was called "Little Hunter". Little Hunter was approximately three years older than Charlotte Rose. They often played and frolicked with each other.

Little Hunter graduated from high school during the buildup of the United States military forces of World War One. After President Woodrow Wilson had declared war against Germany in early April of 1917, Little Hunter enlisted into the US Army in late August of 1917. He had celebrated his eighteenth birthday on April 1, 1917. Little Hunter was assigned to Fort Leonard Wood, Missouri for his basic training. He was being trained to be an infantryman when he was killed on a training assignment. Little Hunter died on May 15, 1918. He was 19.

Funeral services for Hunter Smithfield Jr. were conducted on the last Saturday of May 1918. Charlotte Rose attended the home going celebration, carrying in her arms her three-month-old daughter, Rose.

Charlotte Rose, who identified as a Creole, and her infant daughter, were the only nonwhite to attend the funeral at the white segregated Mule Ridge Baptist Church. She, carrying baby Rose, had walked approximately three miles to be present. It was the first time that Mrs. Adaline had the opportunity to meet the newborn. When Mrs. Adaline gazed at baby Rose, she too pondered in her heart, like Mabel, that her son was surely the father. Baby Rose had inherited from her father his eyes, nose, and hair that were exactly as she remembered Little Hunter having when he was an infant.

After meeting and holding baby Rose for a few minutes, Mrs. Adaline invited Charlotte Rose and her infant to sit with the immediate members of the family on the front row pew. Mr. Hunter Smithfield Sr. was in total agreement that Charlotte Rose and her infant should sit with the family. Charlotte Rose sat silently through the entire eulogy with tears streaming from her eyes.

On Sunday, the day after the home going celebration for Little Hunter, Mabel and Mrs. Adaline began to reminisce concerning the time that Charlotte Rose and Little Hunter had shared together playing and frolicking. They had been with each other every day that Mabel worked in the home of Mrs. Adaline. Mr. and Mrs. Smithfield had allowed Mabel to bring Charlotte Rose to work while Mabel attended to her chores.

"After they became teenagers, we should have forbidden them from spending so much time together," they both said in unison with disbelief.

It was just after midnight, Saturday morning, exactly one week following the funeral of Little Hunter, when Charlotte Rose wrote a note and left it in the crib with her infant baby, Rose. The note read:

Dear Mom and Dad,

I am so sorry for getting pregnant. Thanks for supporting me through my pregnancy. I made a mistake, and it has caused each of you much, much embarrassment. I AM SORRY! PLEASE FORGIVE ME! I know you will love and take care of Rose. I have taken the monies that you all were saving in the cookie-jar bank. I will be leaving the Smithfield Plantation. Give my thanks to Mrs. Adaline and Mr. Hunter. They were always so kind to me. They allowed me to attend school with Little Hunter and his friends. PLEASE DO NOT TRY TO FIND ME! I will survive.

Always my love, Charlotte Rose.

Upon hearing from Mabel that Charlotte Rose had decided to pursue a life away from the Smithfield Plantation, Mr. and Mrs. Smithfield made a sacred promise to Andrew and Mabel that they would use their enormous financial resources, as well as their political connections, to locate and bring Charlotte Rose back home and to the Smithfield Plantation.

Mr. and Mrs. Smithfield also employed the services of a construction and home builder contractor to renovate and remodel the dwelling place of Andrew and Mabel. The new construction would include brick remodeling, with indoor plumbing, and propane gas for heating and cooking. Andrew, Mabel, and Rose would live with comfort and conveniences.

Andrew and Mabel continued their employment with Mr. and Mrs. Smithfield. Mrs. Adaline allowed Mabel to bring baby Rose to work with her, just as she had done with Charlotte Rose.

As the time and years passed by, Mrs. Adaline grew more and more attached to Rose. When Rose began to walk and talk, Mrs. Adaline began to home school her. Home schooling for Rose continued until she had completed the equivalency of the twelfth grade. Through the years, Rose had spent much of her life living and sleeping in the home of Mr. and Mrs. Smithfield. Mrs. Adaline instructed Rose to address her as "Nanna" and Mr. Smithfield as "Big Daddy" when she was in the privacy of the Smithfield's home, but when she was out in the public, she was to refer to them in the formal address of Mr. Hunter and Mrs. Adaline.

In September 1926, when Rose was twelve years old, her grandparents, Andrew and Mabel Broussard, both were diagnosed with Tuberculosis. Mr. and Mrs. Smithfield made financial arrangements that they would be moved to a sanitorium in St. Louis, Missouri for medical treatment and care. Rose would continue to live under the supervision of Mr. and Mrs. Smithfield.

It was the month of December 1926 when, within a week of each other, Andrew and Mabel died of their disease. Both Andrew and Mabel were in their mid-fifties. Their birth dates had never been officially recorded.

Bearing all financial expenses, Mr. and Mrs. Smithfield arranged for the bodies of Andrew and Mabel to be returned to the Smithfield Plantation for a proper funeral and burial.

On December 16, 1926, a simple graveside service for Andrew and Mabel Broussard was officiated by Reverend Aaron Moses, Pastor of the Mule Ridge Baptist Church. The burial was attended by Rose, their granddaughter, and Mr. and Mrs. Smithfield. Pastor Moses opened the graveside ceremony with the reading of Psalm Number 23, and closed with Ecclesiastes, the third chapter.

Andrew and Mabel Broussard were laid to rest in side-by-side graves in the Negro section of the Mule Ridge Cemetery. They lived their life side-by-side, so it was fitting that they were buried likewise.

Along with the grief and mourning, Rose and the Smithfield's were disappointed and saddened that Charlotte Rose had not attended her parents' funeral. The Smithfield's had published funeral notices in the St. Louis, Missouri, Chicago, Illinois, and Los Angeles, California newspapers expecting—based on private investigation reports—that Charlotte Rose had made it to one of these cities.

When Alonzo was approximately fourteen years old, he and his mother, Leah, were working in the fields chopping cotton. It was a hot summer day in late June when a thunderstorm developed and began to pour rain. Seeking shelter from the rain, the cotton choppers, numbering approximately twenty females and twenty males, sought shelter beneath a large oak tree. A lightning bolt struck the oak tree and the workers. Leah and two male cotton choppers were killed. Leah died while Alonzo, her son, held her in his arms.

Fear, hurt, and sadness gripped and consumed Alonzo. The broken son of a cotton chopper made a promise to his deceased mother, and to his God, that he would soon leave the Gilmore Plantation, and find his way to the state of Illinois. The purported "Land of Lincoln and Freedom."

Leah Gilmore was buried in an unmarked grave of the Gilmore's slave cemetery. She was given a graveside funeral. Leah's home going celebration was attended by her son, Alonzo, plus Benjamin and James, and their respective mothers. These three boys who were cotton choppers, and present during the lightning strike, were believed to be sired by the same white man, who was known only as the "Field Boss".

The date of Leah's funeral was held on July 3, 1918. Her age was estimated to be thirty. Leah's birth date had never been officially recorded.

On July 4th, a day after the burial of Leah and the two deceased male cotton choppers, all the cotton choppers and nonessential laborers were given the day off from their work so they could celebrate the day of emancipation.

The celebration of the day of emancipation was a God send of relief that engulfed and consumed the emotional, and mental state of the celebrants, who were still grieving the tragic death of Leah, and the two male cotton choppers. It was a bright and sunny day, with a temperature of approximately 79 degrees Fahrenheit, with a humidity reading of 52%, and a clear blue sky. The humidity was especially low for a day in early July.

There was a high-level of boisterous joy that accompanied the happy celebration, which was attended by approximately seventy-five cotton choppers and farm workers, including children.

The preparation of the food had begun well before sunrise with the barbecuing of two pigs, weighing approximately 125 pounds each, and a young goat that weighed approximately 50 pounds. The pigs and goat were being cooked over an in-ground BBQ pit that was fired with seasoned hickory wood. This preparation and cooking of the pork and goat were attended to by skilled cooks. The food was plentiful, and in addition to the BBQ pigs, which would be the main course, there was fried chicken, fried fish, hot dogs, and various side dishes of potato salad, baked beans, homemade rolls, cornbread, cakes and pies, and homemade ice cream. The chore of freezing the ice cream was assigned to the children, who took turns at supplying the ice, and turning the mechanism that froze the recipe into ice cream. Special, and unidentified men supplied an abundance of home-brewed whiskey and wine.

Activities consisted of sack races for the children. Horseshoe pitching was enjoyed by all ages, with both males and females participating. Many of the teen aged boys and young men participated in a game of baseball.

The annual Fourth of July celebration was financed by the Gilmore family, who had constructed this park for this occasion and other special events.

Alonzo used the celebration of the emancipation to summon Benjamin and James to make plans to escape from the Gilmore Plantation. While they hated the environment of the plantation, they hated their alleged father, Field Boss, even more. The Field Boss's responsibilities were to provide for safe work environments, safe living quarters, and to report all medical needs to Mr. Gilmore, the plantation owner. Instead of carrying out these prescribed responsibilities, the Field Boss used his position to intimidate and frighten the men, and abuse and rape the young girls and women.

Using the atmosphere of the festivities, the smell of meat being cooked, the laughter, and loud talk and conversations, which lasted late into the evening, as a distraction, The Boys plotted their plan to run away from the plantation. The Boys, who were approximately the same age, with Alonzo being the oldest, shared similar physical characteristics that included light complexions, light grey eyes, and curly black hair. They ranged in weight from about 130 to 150 pounds. Alonzo was the heavier of the three.

The plot that The Boys discussed received input from James, the youngest of the three. "Why don't we just kill that bastard, the Field Boss? Just as we know that he is our father, he, too, knows that he forced our mothers to have sex with him."

"No, we can't kill him," Benjamin offered. "The police will know that we killed him, and then they will hunt, and lynch us," concluded Benjamin.

Alonzo decided, "We must slip away around midnight. This time would allow for the adults to be asleep after the ample amounts of food and alcohol that they had consumed during the Fourth of July celebration. Benjamin and James, share the plan with your mothers, and seek their blessings and prayers."

When James shared the plans with his mother, Flora, that he and his half-brothers had made, she blessed her son. And with tears in her eyes, she prepared a change of underwear, socks, and a second pair of overalls, work clothing, and a pair of work shoes, along with a two-day supply of canned meats and fruit. All stuffed into a burlap bag. "James, do not forget to pray, and always ask God for guidance, and His grace and mercy. And, I want you and your brothers to stay together, and always support each other." James's mother tearfully embraced her son, and said goodbye.

Benjamin's mother, Ada, was not surprised when her son shared the plans that he and his brothers, James and Alonzo, had made at the grave site where they said goodbye to Alonzo's mother and placed a wood grave marker at the head of her grave. The grave marker read "DEAR MOM, LEAH GILMORE". The grave marker had been prepared by one of the negro female housekeepers, who had been taught how to read and write. The Boys were never schooled and did not know how to read or write.

"I knew that you boys would soon be leaving the Gilmore Plantation," began Benjamin's mother. "I had always kept a bag of clothing packed for you, and now that Leah is gone to be with the Lord, I will prepare a bag for Alonzo," Ada concluded, with a hurtful sigh in the tone of her voice, as she advised Benjamin to stay with his brothers, and to let God guide his steps.

The Boys' plan called for them to walk north, staying close to the main roads and highways, but staying a safe distance into the wood lines. This strategy for foot travel would provide a measure of safety to avoid being captured by bounty hunters, and members of the Ku Klux Klan (KKK), who were capturing runaway plantation workers and returning them back to their plantation owner or forcing them to join the US Military. These endeavors were done for profit. The Boys' plan also called for them to search out negro farmers or white farmers who needed cheap laborers and were empathetic to their plight and dreams. The Boys' farm and labor skills included the milking of cows, and cleaning of livestock barns and chicken houses. The Boys were also adept at the killing, butchering, and curing of beef and pork. And in exchange for their labor, The Boys requested food, a place to wash their clothes, and a dry place to sleep, with bed blankets or other covering to keep them warm in the cold weather. The Boys usually slept in a livestock barn or animal food shed. This part of their plan called for them to stay with their host for three to five days, and then continue their journey through the summer and fall months, but during the winter, they would hunker down for the entire season.

During the second year of their journey, The Boys believed they had traveled through the state of Tennessee and were somewhere in the state of Arkansas. While wintering over on a farm in Arkansas, James contracted pneumonia in both of his lungs. Fearing that they would be captured by bounty hunters or the KKK, The Boys chose not to seek medical help, but attempted to treat James's illness themselves. Their well-intentioned efforts failed, and James died of his disease during the spring of 1919. Alonzo and Benjamin also hid James's illness from their host farmers, who were helping them to survive the winter months. After James's death, the remaining two Boys went to the host farmer, and informed the patriarch of the negro family what had transpired. Securing a promise of silence from the gentleman, Alonzo borrowed a shovel from the host farmer, where he and Benjamin found a clearing deep into the woods, with a large oak tree that was approximately twenty-feet in height, with huge limbs and with branches which had begun to bloom and leaf for the upcoming spring and summer season. Alonzo and Benjamin by taking turns, dug a very deep grave for James. They were sure that the depth of the grave would prevent dogs and other wild animals from disturbing the remains of James. As neatly and with meticulous care and details, Alonzo and Benjamin arranged the overall and red-plaid shirt, while with their fingers smoothed his straight-black hair to the back of his head. After removing James' bull-horn-handle pocket Case Knife from his right front overall pocket (each boy had been gifted, by older men on the plantation, with a Case pocket knife. At the oak tree, which was bustling with birds and squirrels building and preparing their nests for a new season of life, with prayers and sadness, Alonzo and Benjamin, taking turns with their brother's Case knife, carved three Christian crosses into the trunk of that strong oak tree, where, with his pocket knife, they buried James, with his feet pointing south and his head north somewhere in the State they believed to be Arkansas.

"Alonzo, do we say our good-bye to James"? Benjamin asked quietly as he spoke in a quivering and tearful voice.

"No", exclaimed Alonzo! With his right arm around his brother's shoulder, "We will just say farewell", while he quietly hoped and prayed that he and Benjamin would one day return and place a permanent head stone at their brother's grave!

It was late April or early May when now the two boys continued their journey to reach the state of Illinois, the Land of Lincoln and Freedom. After swimming a second river, which The Boys calculated they would have to swim to reach the state of Illinois, Alonzo and Benjamin believed that they had arrived at their planned destination. It was late fall, their third winter was nearing, and they had to find a farm so they could hunker down for the winter months. The Boys believed the year to be late 1919.

Chapter Four

The Mule Ridge Plantation

ALONZO AND ROSE CONTINUE TO REMINSCE

M oving out from the cover of the densely wooded forest, the first building that Alonzo and Benjamin were to see was a pristine white structure. The Boys recognized the structure to be a place of worship. The edifice had a Christian cross at the front and on the top of the building. The Boys noticed how neat and well maintained the surrounding grounds were. The grass lawn was neatly mowed and manicured, and the shrubbery and trees were professionally pruned. There were three additional buildings adjacent to the church edifice. One of the buildings appeared to be the home for someone. Walking up closer to the buildings, Alonzo and Benjamin saw a male figure enter through the front door of the church building.

After knocking on the front door of the building, The Boys were warmly greeted by an imposing figure of a man. The gentleman was of a dark brown complexion and had comforting, but piercing dark brown eyes. His hair had the appearance of sheep wool, except the hair was coal-black in color. This giant of a man stood about 6 feet 3 inches and weighed approximately 200 pounds. With large, strong, but gentle hands, the gentleman shook hands with The Boys and introduced himself to be Reverend Alonzo Moses, Pastor of the Mule Ridge Baptist Church. "And whom am I in the company of?" the pastor asked in a deep and comforting baritone of a voice.

"I am Alonzo, and this is my brother Benjamin," Alonzo replied in a controlled but nervous voice.

Noticing that The Boys were unshaven and unkempt, and that their clothing and shoes were extremely worn and tattered, Pastor Moses stated the obvious. "You Boys must be hungry. Come in. Please, join my wife and I for the big meal of the day," the pastor said as he held the door open and pointed his hand inside, palm up. It was approximately three o'clock in the afternoon.

The pastor walked Alonzo and Benjamin down a hall and into the dining room of the pastor's quarters. "This is Mrs. Angelica Moses. I call her 'Angel'. She will insist that each of you call her 'Mother Moses'. That is how everyone living on the Smithfield Plantation addresses her. Plus, she is the mother of the Mule Ridge Church," said the loving husband, as he completed the introduction of his wife.

Mother Moses presented an imposing and confident stature of a woman. She was a stout-built lady, standing approximately 5 feet 8 inches, and weighing an estimated 170 pounds. Her complexion was light brown skin, framed with expressive, dark brown, large eyes, and a broad nose. Mother Moses wore her coal-black straight hair pulled to the back and into a bun.

While Mother Moses was preparing the dinner table, The Boys were instructed to go out onto a covered and screened in porch that was used to wash one's hands and face before each meal. Soap and towels were provided for this routine.

Before dinning on an ample amount of vegetable soup, mashed potatoes, baked chicken, collard greens, and cornbread, Pastor Moses offered a prayer for the blessing of the food that God provided for nourishment, healing, and health. While each one was holding hands with the next, everyone said, "Amen."

After finishing with the main course meal, Mother Moses offered to her diners a serving of peach cobbler pie and milk. It was at this time that Alonzo and Benjamin began to feel at ease and trusting of the Pastor and Mother Moses, and The Boys emotionally poured out their life's story and journey from Tupelo, Mississippi, and the Gilmore Plantation. They confessed that while Gilmore was their last name, each of their mothers were negro and they were fathered by the same white man, knowing him only by the name of "Field Boss".

After preparing dinner plates filled with leftovers from the big meal, Pastor and Mother Moses escorted Alonzo and Benjamin to the migrant hut, where they would live during the winter months. Pastor Moses used this hut to house migrants and other seasonal workers as they were migrating to the northern cities in search of jobs and a better life.

The hut where Alonzo and Benjamin would be hunkering down during the winter months was about fifty yards from the home of Pastor and Mother Moses. The hut was equipped with kitchen, living room, and bedroom furniture. The kitchen was equipped with a two-eyed cook stove and contained a variety of cooking utensils, with an adequate assortment of plates, saucers, cups, knives, forks, and spoons with a dining room table with four chairs. The room that was both living room and bedroom was equipped with a wood burning heating unit, as was the cook stove. The furniture consisted of bunk beds, a couch with end tables, two comfortable chairs, and a chifforobe having a compartment for hanging clothes and drawers for folded clothes, underwear, and other small items of clothing.

The hut was of wood frame construction. The exterior walls and roof were covered with a tan-colored sand and tar material that was manufactured to resemble bricks. The interior walls were finished with a tongue-and-groove white pine wood. The dimensions of the huts were approximately 25 feet in length and 12 feet in width. Each hut had a two-step entrance through the northeast facing living room/bedroom, solid-wood front door, with eight-pane windows that could slide open up and down, one in the front and one on each side. There was an identical back door entrance through a southwest facing screen door that led to a screened-in, covered porch, which measured 8 by 5 feet. The kitchen had one eight pane window and the door that led into the porch. The huts were well insulated with a fiber-glass-like material. The craftsmanship of the huts was done with a professional touch and with perfection. The Boys had agreed to stay and work on the farm in exchange for food and shelter during the winter season.

Before leaving the company of The Boys, Mother Moses removed a tape measure from the pocket of her apron, which she always wore, to measure The Boys' heights, arm lengths, waist sizes, and their feet lengths and widths. Alonzo and Benjamin each measured 6 feet 1 inch in height and weighed approximately 130 pounds. During their arduous journey, Alonzo and Benjamin had lost a considerable amount of weight but had grown in height.

Holding hands, while walking slowly back to their home, Pastor and Mother Moses were overtaken with an emotional and heavy heart concerning the ordeals The Boys had endured during their short life. Not having been blessed with any biological children of their own,

Pastor and Mother Moses pondered in their hearts how they could legally adopt The Boys.

The next day, Mother Moses would travel to the town of Wyatt for the purpose of purchasing clothes, shoes, and some fresh meat and other grocery items for The Boys. She also planned on scheduling Dr. Fairweather to travel to the church to perform a physical examination on Alonzo and Benjamin.

Feeling especially blessed and thanking God for their good fortunes, Alonzo and Benjamin solemnly remembered their brother, James. With the safety and comfort of the hut, The Boys were sure that they had arrived somewhere in the state of Illinois.

On their journey, Alonzo and Benjamin had been bathing themselves from creeks of water, or from watering troughs used to water farm animals. The Boys began to make fires in the heating units, a cook stove in the kitchen, and a heater in the living/bedroom area. The hut was well supplied with fuel wood. Utilizing two five-gallon containers to heat the water from a hand pump well, they were about to have their first hot bath in two years. The Boys secured a six by three-foot metal bathtub that was stored on the covered, screened-in back porch.

After washing their tattered and torn clothes, and taking a hot bath, The Boys ate the plates of food that Mother Moses had prepared for them to take to their living quarters. Not having slept in a bed with adequate and warm coverings since they left the plantation, The Boys, by the light of four kerosene lamps, once again thanked God for keeping them safe.

With full stomachs, Alonzo and Benjamin enjoyed a comfortable and restful night of sleep. Waking early in the morning, The Boys discovered that the food pantry was stocked with a variety of non-perishable food. After preparing and eating a breakfast that consisted of hot oatmeal and canned peaches, The Boys arrived at the home of

Pastor Moses at approximately 6:30 a.m., just before sun rise. Being an early riser himself, Pastor Moses was enroute to the church's sanctuary for his daily hour of prayer and meditation.

"Good morning," Alonzo and Benjamin said in unison as they cheerfully greeted Pastor Moses. "We are prepared for work. What do you need for us to do today?" The Boys humbly inquired.

"First, let us go into the sanctuary for a word of prayer and thanks to our Lord and Savior Jesus," replied Pastor Moses.

"Yes sir," The Boys responded, with wide smiles and heads held high.

"Then young men, the two of you can go with me, and we can harness Angel's mule, whom she has named Samuel, and prepare her buggy as she plans on going into town today so that she can do some shopping," Pastor Moses instructed the young men.

With the help of Alonzo and Benjamin, Pastor Moses soon completed the task of harnessing and hitching Samuel to Mother Moses's favorite buggy. Pastor Moses then instructed Benjamin to deliver the mule and the buggy to Mother Moses, who was awaiting her transportation, so that she could travel to town. When Benjamin returned to the stable, Pastor Moses and Alonzo had harnessed and hitched a team of two mules to a wagon, which Pastor Moses would use to tour some of the operations that were performed on the Smithfield Plantation. The operations that Pastor Moses pointed out to Alonzo and Benjamin, included but were not limited to the following:

(1). A forty-to-fifty-acre lake which was called the "Blue Hole". The lake of water was naturally formed by the forces of Mother Nature, and no one knew how deep the Blue Hole was. The source of water which supplied the lake derived primarily from underground springs of water. This constant source of underground spring water kept the water in the lake moving, and in combination with its depth, the lake never experienced a hard freeze. Usage of the lake included watering of the farm animals, swimming, baptism of believers into the Christian faith, and other recreational activities by the residents of the Smithfield Plantation. The north end of the Blue Hole was fenced off for usage by the white population, and the south end was fenced off for the negro population. The fencing also served to keep the farm animals from entering into the recreational sections of the lake. To keep the Blue Hole clean, healthy, and safe was a priority, as it contained a large population of fish consisting of several species. The fish were a major source of food for the residents of the Smithfield Plantation.

Mr. Lonnie Wallace, and his three sons, supplied most of the fish for the residents of the Smithfield Plantation. The Wallace family were responsible for maintaining the health and safety of the Blue Hole. A major percentage of the Wallace's' income was derived from the sale of fish, not only to the Smithfield residents but to neighboring villagers as well. In addition to harvesting fish from the Blue Hole, the Wallace family also harvested fish from the Mississippi River. The harvested fish were kept alive and healthy by keeping the fish in chicken-wire constructed cages that were submerged in the water of the Blue Hole. The species of fish included but were not limited to: buffalo, carp, blue gill, and catfish.

The Wallace family was one of the few families who were contracted with the Smithfield Plantation to rent the land that they farmed. Renters' contracts called for the renter to provide labor, seeds, and whatever equipment was needed to produce a crop for delivery to market. The renter would pay the Smithfield Plantation twenty percent of the net income, while keeping eighty percent to finance their farming operation.

The sharecroppers' contract with the Smithfield Plantation called for the sharecroppers to provide one hundred percent of the labor, seeds, and equipment for one third of the net profit. There were no negro renters contracted with the Smithfield Plantation.

(2). The Smithfield mule breeding operation was owned and managed by the Smithfield family. Mules were essential in the successful and profitable operation of the Smithfield Plantation. The Mule Ridge village received its name and reputation from the breeding practices utilized by the Smithfield mule breeding business. The breeding application employed the mating of an American Mammoth male donkey, a very large breed of donkeys, with a female Irish Draught horse. This female horse was a large breed of horses and was usually black in color. This mating would produce a large mule that had strength, endurance, and gentleness. These offspring, which were hybrid and sterile, would sell for $150-200 each and was a substantial source of income for the Smithfield family. The number of mules sold annually throughout the Cotton Belt was estimated at 1,500 to 2,000.

During the period of slavery and through the mid-1900s, mules were the equivalency of modern-day mechanized farm equipment. Mules were used to plow and till the soil for planting farm crops, including cotton, corn, alfalfa, and other grain crops, to transport wagon loads of cotton to the cotton gin, to harvest and haul trees to the sawmills to produce lumber. And, mules were used for basic human transportation, and other recreational activities. All of which fueled the economy of the United States of America.

(3). The Smithfield hog and beef operation annually produced approximately five thousand hogs, and two thousand heads of cattle, which were sold throughout the southern states. This operation employed fifteen to twenty laborers and was a major source of income for the Smithfield family.

(4). The logging and lumber operation employed five to ten fulltime employees, who annually harvested three to five hundred 20-30-foot-tall mature oak and white pine trees.

An estimated fifty percent of the lumber was processed for the construction of housing and living quarters for the renters, sharecroppers, and migrant laborers who populated the Smithfield Plantation. There were an estimated seventy-five of these buildings consisting of one to three bedrooms each. Another twenty to thirty percent of the lumber was used for the construction of animal barns, feed sheds, and posts for fencing. The remaining lumber was sold for profit. Approximately a thousand acres of the ten-thousand-acre Smithfield Plantation was wooded.

As their faces were caressed by the coolness of the October breeze, Pastor Moses quietly interrupted the silence. "Young men," he said, thanking God for allowing each of them to witness the beauty of this fall day which presented a high blue sky, and a panorama of colors consisting of brown, orange, yellow, and red leaves that clothed the trees and bushes. "Alonzo, Benjamin," said Pastor Moses to The Boys in a fatherly tone of advice, "you can work at one of the Smithfield operations, which will require you to work five full days a week, and a half day on Saturday with pay for a full day at $1.50 per day, or you can go and work for one of the renters. Working for the renters will be part time only, and the pay will be $1.25 per day. I advise each of you to work for Mr. Smithfield.

"Renters and sharecroppers grow approximately two thousand acres of cotton. You young men can hire yourselves out to chop and pick cotton, and work the corn fields, which need to be weeded and harvested." Pastor Moses placed his right thumb and forefinger on his chin. "Matter of fact, there is still some cotton that need to be picked, and some corn that need to be harvested. Young men, you can hire yourselves out as early as tomorrow to do that type of work. Also, there are select and unidentified gentlemen who labor in the production of alcoholic beverages, moonshine," said Pastor Moses with a derisive and unapproving tone in his voice.

Alonzo and Benjamin shook their heads vigorously, and replied in unison with a loud, "No!"

"Sir, we have been chopping and picking cotton and doing that type of work all of our life," Alonzo said in a strong and definitive voice.

Appealing compassionately, Pastor Moses encouraged Alonzo and Benjamin to volunteer their time and skills to help the sharecroppers and renters with the butchering and curing of their pork for the coming winter months. Some of the pork was cured by slowly smoking the fresh slaughtered meat over hickory firewood that was performed in a shed with a dirt pit. The shed was called a "smoke house". With the smoke venting through small cracks in the roof of the building This process required the temperature to be maintained at 180 to 200 degrees Fahrenheit for twenty-four to forty-eight hours and required careful and in-person attention. The skin and fat tissue of the butchered hog was cut into one- to two-inch squares and rectangles for the purpose of producing lard, which was used to fry certain foods, including but not limited to chicken, fish, rabbit, and other small animals. The production of lard was accomplished by placing the cut pieces of skin and fat tissue into five- and ten-gallon iron pots and cooking for four to eight hours over a hot wood fire, which produced pork rinds, called cracklings. The remainder of the meat was cured with a specially formulated salt.

Chitterlings (hog intestines), and head cheese were carefully cleaned and processed, and when seasoned and cooked properly were considered a delicacy. Head cheese was produced from using selected portions of a hog's head, snout, ears, and tail. Hair from the hogs was collected, cleaned, and processed to be used as insulation in the production of quilts and pillows, which was used for human bedding. The hogs' hooves were used in a specially brewed sassafras tea and was purported to prevent respiratory diseases. Chitterlings, head cheese, and surplus hogs' hair were sold by the sharecroppers and renters to supplement their income. The sharecroppers and renters used all parts of the butchered hog for food or monetary profit. With special processing and recipes, all of the hogs' organs were made edible. The young males that were not used for breeding were castrated at approximately three months old, and their testicles, when processed and cooked properly, were claimed to be a delicacy. There was a well-known saying in the community that "The only thing that was lost from butchering a hog, was its 'oink.'"

The process of smoking and salting pork allowed the winter cold weather to assist in its preservation, as the meat was stored in outdoor sheds.

The butchering of beef, which mostly involved male animals, weighing two to three hundred pounds was performed when several families had a need for beef in their diet. The meat was consumed in two to three days due to the difficulty of storing and preserving beef products.

The hides and hair from the slaughtered beef animals were cleaned and processed to be sold to buyers to produce commercial products, such as shoes, gloves, leather coats, and insulation and padding. The sale of these products added to the income of the sharecroppers and renters.

For a healthy diet, the sharecroppers and renters farmed their poultry needs, and received their dairy products from a family-owned milk cow. Canning their vegetable requirements from family grown vegetable gardens, as well as canning their fruit requirements from the Smithfield peach, apple, and pear orchards. Additional nutritional needs were supplied from nut groves of pecans, hickory nuts, and black walnuts. Residents of the Smithfield Plantation enjoyed a healthy and well-balanced diet. To supplement and give variety to their diet, many of the residents of the Smithfield Plantation hunted and trapped wild animals, which included rabbits, squirrels, and raccoons. The residents also killed various species of wild fowl, which included quail, geese, ducks, and turkeys. The furs from the small animals were specially processed and sold to furriers for a profit.

Several of the sharecroppers and renters produced and processed honey and sorghum molasses to supplement their diet, and add to their cash income, as they sold the surplus.

"Young men, if you will bless these sharecroppers and renters with your time and skills to help with the butchering and curing of their winter meat, they do not have cash money to pay the each of you, but they will reward you with canned fruit, fresh meat, and baked goods of cakes and pies. And God as my witness, you will always be invited into their homes for Sunday and holiday meals and other festivities," Pastor Moses promised as he concluded with his advice.

It was approximately three o'clock in the afternoon when Pastor Moses and The Boys returned from touring the Smithfield Plantation. Mother Moses had also returned from her shopping trip, and had left her mule, Samuel, and buggy tied to the hitching post that was just outside of the back porch of her home. Pastor Moses then instructed Alonzo and Benjamin to return both his team of mules and wagon, as well as his wife's mule and buggy to the barn and unharness, feed, and water the animals, before returning to his home where they were invited to share a meal with him and Mother Moses.

Upon returning to the Moses home, Mother Moses instructed and guided The Boys into the living room, where she informed Alonzo and Benjamin why she had gone shopping. Mother Moses had purchased for each of The Boys two pairs of work shoes, three pairs of wool winter socks, three pairs of winter under wear (long johns, as they were affectionately called), three pairs of summer—top and bottom—under wear, three winter shirts in the colors of solid grey, black and red plaid, and green and black plaid, two pairs of overalls, and a pair of blue jeans, and a winter coat. Mother Moses noted that the gifts were financed from the Mule Ridge Baptist Church benevolent funds.

Speaking in cracking voices, Alonzo and Benjamin were finally able to say in unison, "Thanks!" as they embraced and hugged Mother Moses.

Alonzo immediately followed their thanks with a bashful acknowledgement. "We do not have any money to pay for these gifts, but we can work to pay for these shoes and clothes."

"The gifts are from the loving hearts of the members of the church, and gifts are free and unconditional," Mother Moses interjected, while holding the palms of her hands over her heart.

Back to the kitchen and at the dinner table, Mother Moses had yet another surprise for her dinner guests. In addition to the vegetable soup, which she had left on the cook stove to slowly simmer, Mother Moses purchased some lunch meats, consisting of bologna and head cheese, and some bottled soft drinks, in the flavors of orange, grape, and cola. She also purchased saltine crackers, and a loaf of white bread. For dessert, there was a large package of a dozen cinnamon rolls.

After an enjoyable after noon meal, Mother Moses brought out her hair-cutting and shaving equipment. In a commanding but jovial and loving voice, Mother Moses suggested to Alonzo and Benjamin that they "need a haircut and shave!" Barbering was among the many services that Mother Moses provided to the sharecroppers and renters living on the Smithfield Plantation.

With much laughter, Alonzo and Benjamin nodded their heads. "We have not had our hair cut in over two years, and we have never had a shave," Benjamin admitted with laughter.

With neat haircuts and clean shaven, Alonzo and Benjamin collected their gifts of clothing and shoes and prepared to return to their hut, when Mother Moses informed them with more good news.

"On tomorrow, Dr. Fairweather, M.D., and his daughter, Ms. Florence, a registered nurse, will be coming to the church to perform a physical examination on the each of you. They are scheduled to arrive at approximately ten o'clock tomorrow morning. Dr. Fairweather has provided for the each of you two containers. The small container is for urine, and the large container is for a stool sample, *number two*." She instructed The Boys to write their names on their individual containers, and place the containers into the paper bags that were provided.

Hearing the instructions from Mother Moses to write their names on the assigned containers, Benjamin held his hands behind his back and looked at his feet. "Well . . . um—"

"Um . . . ma'am," Alonzo interrupted, "neither of us has learned letters. We do not know how to write our names." He blushed and gave a forced smile.

At this moment of the discussion, Pastor Moses was entering the room from gathering eggs from his chicken house and heard The Boys admission of their illiteracy. Gracefully and respectfully interjecting himself into the conversation, Pastor Moses with Christ-like compassion informed Alonzo and Benjamin, "On the upcoming Monday, which is the last Monday in the month of October, when cotton picking vacation will end, a new school semester will begin. In addition to teaching the regular classes from grade one through eight, I also teach an evening class for adults. The class is conducted on Mondays and Thursdays, from 6:00 to 8:00 p.m. The classes are specially designed to teach adults how to read and write. And I assure each of you that I will teach you how to read and write," exclaimed Pastor Moses.

"Oh, by the way," he continued, "tomorrow is Thursday, the day for the 'Ice Man' to deliver ice. He delivers ice on Mondays and Thursdays and will supply the ice box that is on the back porch of the hut with a fifty-pound block of ice." Pastor Moses presented The Boys with a dozen fresh eggs, and a pound of smoked ham. "I expect to see the each of you at the adult reading, writing, and arithmetic class on Monday at 6:00 p.m. sharp," Pastor Moses reminded Alonzo and Benjamin before encouraging them to have a good night of rest and sleep.

Entering their hut, which The Boys now referred to as their home, Alonzo and Benjamin began to quietly weep with teary eyes and joyfully thanked God for the many blessings that He was bestowing upon them.

Life had been very difficult for the few short years of Alonzo and Benjamin, who were now accepting the responsibilities of becoming contributing, independent young men. They could now see the proverbial light at the end of the tunnel, and the light was not out, but shinning bright, with hope.

While heating water for the purpose of taking a bath, Alonzo and Benjamin happily laid out the new clothes and shoes that they would be wearing for their physical examinations, which were to be administered on the following morning by Dr. Fairweather and his daughter Florence. It was the first time that The Boys had worn new clothes and shoes. While on the Gilmore Plantation, Alonzo and Benjamin had only worn shoes and clothes that were hand-me-downs from older and larger boys.

Awakening early from a relaxing night of rest and sleep, Alonzo and Benjamin prepared breakfast for themselves, which consisted for each of them, two scrambled eggs, two slices of ham, and a bowl of grits, with hot sassafras tea, a beverage brewed from the dried roots of the sassafras tree.

Continuing with their morning routine of meeting with Pastor Moses for prayer and meditation in the church's sanctuary, Alonzo and Benjamin informed Pastor Moses where they preferred to work. Alonzo's choice was to work for the mule breeding operation, while Benjamin chose to work for the hog and cattle operation.

"Very good! I know that you young men will work hard, and do your best each and every day, and remain truthful and loyal to your word and promise," Pastor Moses replied, as he blessed Alonzo and Benjamin, and that he would inform Mr. Smithfield to enter them on to the payroll at $1.50 per day.

"One additional request, before you young men meet with Dr. Fairweather for your physical examination, just as the animal stalls and chicken houses need cleaning and the removal of their body waste, so do the outhouses. I am requesting for you men to help me with cleaning the church's outhouses, both women and men, as well as the outhouse for my home, and of course the outhouse for your home."

"Yes sir, Pastor," Alonzo and Benjamin, in unison, responded.

With detailed instructions from Pastor Moses, Alonzo and Benjamin were to remove the buckets which caught the human waste, and with a mule team and wagon, transport the waste to a cesspool, which was in the wooded area just off the road that headed east. The cesspool would be in the south portion of the woods, and the distance was about a half-mile. With its' smell, you couldn't miss it.

The cesspool was fenced in with eight-foot-high pine boards, and a pine board roof. The cesspool had been designed and constructed with a six-foot-wide and eight-foot-deep band of rocks, gravel, and sand to liquefy, and filter the waste to the surrounding soil. The dimension of the cesspool was 12 feet by 12 feet and was properly vented. The cesspool had been inspected and certified to be safe by the county health department. Pastor Moses further instructed Alonzo and Benjamin to take a ten-pound bag of lime and mix it with the waste before emptying into the cesspool, Adding the lime would help break down the solid waste, speed up liquefication, and control flies and other insects.

The Smithfield Plantation had fifteen of these cesspools strategically located and constructed throughout the housing area of the sharecroppers and renters.

At the conclusion of the instructional meeting with Pastor Moses, Dr. Fairweather and his daughter, Florence, arrived at the church to conduct the physical examinations on Alonzo and Benjamin.

The father and daughter medical team arrived in a 1918 Ford vehicle that was powered by a gasoline engine. The automobile, driven by Dr. Fairweather, was black in color, and specially designed and customized to carry medical equipment.

After warm and nervous introductions, Alonzo and Benjamin presented nurse Florence with the paper bags that contained their urine and stool samples. Alonzo had drawn the letter "A" on his samples, and Benjamin had drawn the letter "B" on his samples. The Boys used the drawings for signatures.

The results of Alonzo and Benjamin's physical examinations turned out to be near identical. Each was 6 feet 1 inch, weighing 130 pounds, with a blood pressure of 118 over 80, and their lungs sounded healthy. A sample of blood was drawn from each for further laboratory analysis.

Because of the low body weight of both Alonzo and Benjamin, Dr Fairweather feared that The Boys were infested with intestinal parasites—worms. With this preliminary diagnosis, and to be on the side of safety, Dr. Fairweather left a prescription of Mebendazole to be taken as one tablet before breakfast and one tablet before bedtime for thirty days. This medication was developed to treat and prevent intestinal parasites. The stool samples upon further analysis would confirm or refute the preliminary diagnosis.

Before scheduling a follow up appointment, which would include a dental examination that would be conducted at Dr. Fairweather's clinic, located in the town of Wyatt, Dr. Fairweather noticed the curiosity and interest that The Boys had in his vehicle, so he offered to give them a ride. Never having ridden in a motorized vehicle, Alonzo and Benjamin gladly accepted.

Once they returned to the church, Dr. Fairweather reminded Mother Moses of Alonzo and Benjamin's next appointment date, and he and his daughter said good-bye then departed.

After the completion of their physical examination, Pastor Moses instructed Alonzo and Benjamin to harness and prepare Mother Moses's mule and buggy, for she had an appointment to visit with a pregnant mother-to-be. Mother Moses was a certified midwife, trained and certified by Dr. Fairweather to aid and assist with the delivery of the newborns of the negro women.

"Be sure to be available that when my wife, Angel, returns from attending to her midwife duties, that one of you will return her mule and buggy to the barn, and water and feed Samuel," Pastor Moses instructed Alonzo and Benjamin, before he reminded The Boys of the Sunday morning worship service.

Chapter Five

A New Day. A New Dawn. A New Life Has Begun!

ALONZO AND ROSE CONTINUE TO REMINISCE

I t was the morning of the last Sunday of October 1919, when Alonzo and Benjamin began their new life. Having never attended a Christian worship service and not knowing what the proper dress should be, Alonzo and Benjamin sat on the outside steps that led into the sanctuary. It was a sunny, but crisp and chilly fall morning. The Boys were wearing their new blue jeans, and their matching grey shirts, with black work shoes. They were dressed like twins, and their physical features were strikingly similar in appearance.

The sermon and lesson that Pastor Moses chose to teach and preach from was Psalm Number 23. (KJV):

1 The Lord is my shepherd I shall not want.

2 He maketh me to lie down in green pastures: he leadeth me beside the still waters.

3 He restoreth my soul: he leadeth me in the paths of righteousness for his name's sake.

4 Yea, though I walk through the valley of the shadow of death, I will fear no evil: for thou art with me; thy rod thy staff they comfort me.

5 Thou preparest a table before me in the presence of mine enemies: thou anointest my head with oil; my cup runneth over.

6 Surely goodness and mercy shall follow me all the days of my life: and I will dwell in the house of the Lord forever.

Using an expository style of teaching the word of God, Pastor Moses revealed that Psalm number 23 taught three very important principles to live "A Purpose Driven Life".

"A Purpose Driven Life, is about relationships: (A). Relationship with family and friends. (B). Relationship of husband and wife, and children, and (C). Relationship with God. These relationships can be defined as: Orientation. Disorientation. Reorientation," Pastor Moses summarized, as he asked the parishioner for an Amen.

In Psalm Number 23, David is in perfect orientation and commitment to God, as exposed in verses 1-3. In verses 4-5, because David committed the sin of adultery with Bathsheba, David finds himself disoriented and separated from God! In verse 6, after confessing and taking responsibility for his sin of adultery with Bathsheba, in Psalm Number 51, David is forgiven and washed clean of his sin, and is reunited with God, and is reoriented in relationship with God," Pastor Moses noted in closing his sermon and dissertation.

The sermon and lesson on Psalm Number 23 left an indelible experience on the minds and in the hearts of Alonzo and Benjamin that they would embrace and carry for the remainder of their lives.

After closing his sermon and lesson with a prayer of benediction, Pastor Moses approached Alonzo and Benjamin, inquiring as to why they had chosen to worship on the outside steps, instead of inside of the sanctuary with the other worshippers.

"We believed that our work clothes were not proper dress for church. The other worshippers were wearing clothing that they do not work in", Alonzo replied, as Benjamin nodded in agreement.

"As long as one's clothing are neat and clean, anyone can come and worship as they are. God looks at the heart, not at what a person is wearing," Pastor Moses said, with a tone of spiritual encouragement in his voice.

On the following Sunday's hour of worship, Alonzo and Benjamin worshipped in clean and neat work clothes, but chose to sit on pews at the back of the sanctuary.

Before departing the presence of The Boys, Pastor Moses noted that tomorrow was Monday, the first day of the new school semester, which began on the last Monday of October, and that Mr. Smithfield would be present to meet and greet the new employees, the students, and their parents. Mr. Smithfield did this for both the negro and white residents. The schools and worship centers on the plantation were segregated.

Time for the meet and greet was scheduled for 8:00 a.m. on Monday morning, inside of the church's fellowship hall, which also served as the negro elementary school.

"Young men, remember to report for work following the meet and greet. And your adult reading, writing, and arithmetic class will begin on Monday at 6:00 p.m.," Pastor Moses reminded Alonzo and Benjamin with a commanding but fatherly tone in his voice.

Promptly at 8:00 a.m. on the last Monday of October 1919, Mr. Hunter Smithfield arrived, driving a black, two-door, 1918 Ford automobile, at the Mule Ridge Baptist Church. Pastor Moses was honored to introduce and present Mr. Smithfield to an estimated gathering of seventy-five residents of the Smithfield Plantation. Those present included new employees, and parents and their children who were beginning a new school semester.

Mr. Smithfield was an overweight middle-aged male, weighing approximately 250 pounds, and standing an estimated 5 feet 8 inches in height. He wore a beard that was gray-white in color, with matching hair. His eyes were blue in color, with a comforting gaze for those in his presence. He reminded one of a happy and jolly Santa Claus. Mr. Smithfield's demeanor and engaging smile encouraged comfort and trust.

Acknowledging the many contributions and sacrifices made by Pastor Moses, which included being the spiritual leader of the negro community, elementary school teacher, and representative spokesperson for the negro community to the Smithfield Plantation, Mr. Smithfield said to the sharecroppers and laborers, "Welcome. It is nice to meet our newer residents and workers here at the Smithfield Plantation. And hello to all the familiar faces. Be assured that you will always have safe and comfortable housing, nutritious food, clothing, and a place to worship God, and educate your children. Be sure to give your request of what kind of meat and or turkey that each family wants for Thanksgiving dinner to Pastor Moses, and I will make sure that the family's choices will be met," Mr. Smithfield promised. He shook hands with everyone present before leaving the meet and greet assembly.

As the assembly dispersed and people went about their daily routines, the sharecroppers and renters began preparing for hog butchering season. Tons of wood had to be prepared and made available for heating water, and smoking fresh meat. The smoke houses had to be cleaned and vented so that the fire and smoke pits would burn safely.

Pastor Moses now had to put on his teacher persona as he prepared the church's fellowship hall to function as a one room school for grades one through eight. The classroom had the capacity to seat thirty students. The teaching model used by Pastor Moses called for the higher-grade students to tutor the lower grades, while Pastor Moses concentrated on teaching the higher grades. For a one room school, this model proved successful. Books and teaching materials were recycled from the segregated white school. Negro schools were never issued new books and teaching materials.

Walking approximately one mile to their work assignments, Alonzo and Benjamin discussed what their life would be if they had been born, reared, and educated on the Smithfield Plantation, instead of the Gilmore Plantation of Tupelo, Mississippi. The Boys were in awe of how well-dressed and groomed the students were, and they seemed so happy and eager to attend school and receive an education.

Arriving at their place for work, Alonzo and Benjamin were eager, but filled with anxiety, to meet their respective supervisors. Alonzo would be working under the supervision of Mr. Amos Beck at the mule breeding operation, while Benjamin would be working under the supervision of Mr. Julius Beck at the cattle and hog production operation. The Becks were brothers.

Alonzo and Benjamin each were given safety and health instructions, and issued work clothing and gloves, which included rubber boots to be worn over their work shoes, and rubber overalls to be worn over their work clothes. These rubber boots, overalls, and gloves were to be worn when in contact with the animals, and removing manure while cleaning the animal stalls.

The manure was transported to a location one mile south of the barns, and unloaded onto an asphalt pallet that measured 100 feet by 150 feet and was enclosed, including the top, by heavy chain-linked fence. The manure was separated according to the type of animal, as the composting required different processes for each species of animal. The composting process required various types of vegetation, lime, and periodic stirring and turning of the manure. The composting yard was open to the sunlight and natural ventilation, which assisted with the composting process. The processed and composted manure was used to fertilize the farm crops.

Close contact with the animals occurred when: (1). The male pigs and male calves were being castrated to be later sold for meat and human consumption. Castration took place when the male animals were two to three months old. One percent of the male animals were retained for the purpose of breeding. (2). Before the breeding season, the breeding males were separated from each other, and from the females until the breeding season. The breeding season was managed and scheduled so that birth of the newborn animals would take place during late spring, and early summer months. Meticulous records were maintained to prevent in-breeding of the animals. (3). Livestock found deceased, due to wild animals or natural causes, could be disposed of only after a licensed veterinarian from the county department of animal disease and control took samples from the carcass to determine if the animal carried a contagious or transmittable disease. The deceased animal was disposed of by burning, or a licensed scavenger-service company. And (4). When loading the animals to be transported for sale and slaughter.

Alonzo and Benjamin were each presented with a mule and saddle with a riding harness, which was to be used for herding the farm animals, inspecting the fences for damages, and recreational use and transportation. The Beck brothers reminded The Boys that workdays began at sunrise and ended with sunset, except on Mondays and Thursdays. On those days, Mr. Smithfield authorized for them to leave work at 3:00 p.m. so they could attend their 6:00 p.m. adult reading, writing, and arithmetic class.

"Welcome to the Smithfield Plantation. Your work assignments for today will be to ride and inspect the fences, because early next week we will be moving the animals to the crop fields that have been harvested. We wish each of you well with your classes," noted Amos Beck, with his brother Julius smiling in agreement as they completed their indoctrination and instructions.

"Thanks so very much for the opportunity. We will do our best and work hard," Alonzo promised, as Benjamin smiled while clapping his hands.

Chapter Six

One Day. One Week. One Month. One Year ... at a Time

Alonzo and Rose Continue to Reminisce

On Monday, Alonzo and Benjamin's workday ended at three o'clock in the afternoon. After feeding, watering, and securing their mules, they half ran, half skipped the one mile to their home to get ready for adult reading, writing, and arithmetic classes.

After taking a sponge bath to ensure good body hygiene, changing out of their work clothes into clothing that would be more acceptable in a classroom, and eating a ham sandwich and canned peaches, Alonzo and Benjamin proceeded to the church's fellowship hall to attend their adult class of reading, writing, and arithmetic. It would be their first time in a formal learning and educational setting. Three additional adults—two females and a male—were present for the class.

"I am Reverend Aaron Moses, pastor, schoolteacher, and spokesperson for the negro residents to Mr. Smithfield, owner of the Smithfield Plantation. Please introduce yourself by giving only your first name. Like myself, I understand that many of us carry the last name of our grandparents, who were given the last name of their slave owners; therefore, we respect your desire to keep that information private," Pastor Moses said to the class. Each student stood up and nervously introduced themself.

"Alonzo."

"Benjamin."

"Sadie."

"Ida."

"Willie."

Pastor Moses continued with the class orientation by presenting each student with two pencils, a "Big Chief" tablet for writing, a set of the alphabet, a set of one hundred word/object association flash cards, a set of numbers 0 through 100, a set of arithmetic symbols consisting of the plus, minus, multiplication, and division, and a new, pocket-sized, black, King James version of the Holy Bible. With detailed verbal instructions on how to memorize the alphabet, and to count from 0 to 100, Pastor Moses promised and encouraged the eager-to-learn students that each of them would be successful in learning the skills to read, write, and do arithmetic. "Be blessed, thank God, and go in peace, and I will see each of you here at 6:00 p.m. Thursday," concluded Pastor Moses.

Arriving at their home clutching their learning and teaching materials, Alonzo and Benjamin began to reminisce about their life and journey away from the Gilmore Plantation, and the sad death and burial of their brother, James, and Alonzo shed a tear for his mother, Leah. Within one month, the life of Alonzo and Benjamin had moved from desperate survival to purpose driven, and filled with faith and hope.

Beginning with the turkey that Mr. Smithfield had promised to each family, Thanksgiving dinner and celebration with Pastor and Mother Moses became an annual event. At the dinner table, Mother Moses reminded The Boys that the following day, at 10:00 a.m., they had been excused from work because they had a follow-up physical examination, and dental examination at the Fairweather Clinic, and noted that Dr. Fenton Armstrong, D.D.S., would be conducting their dental examinations.

At the follow up, Dr. Fairweather informed Alonzo and Benjamin that the laboratory results of their blood and stool samples confirmed the preliminary diagnosis that each of them were infected with intestinal worms, and that he would be prescribing another month's worth of Mebendazole, which would rid their system of the parasites; furthermore, all of their other vitals and body functions were normal, and each of them had gained five pounds of body weight.

Dr. Armstrong's dental examination began with Alonzo, as he made the older lad comfortable by noting, "Age before beauty," with Benjamin vocalizing agreement with a loud outburst of laughter. "Young man, you have three cavities. Now how did that happen?" Dr. Armstrong inquired rhetorically, with a wry smile. "Today I will perform cleaning, and give each of you young men some points on dental hygiene, and on the next appointment, I will fix those cavities," he noted, assuring his young patient that everything would be okay.

"Now is the turn for 'beauty'," Dr. Armstrong announced, as beckoned with an out stretched hand for Benjamin to sit in the examination chair. "You, and your brother not only look like twins, but the each of you have identical cavities," said Dr. Armstrong as he performed the examination and cleaning of Benjamin's teeth.

Noting that The Boys had never received professional dental care, and that their oral hygiene practices were poor, Dr. Armstrong told them that their teeth and gums were surprisingly healthy. Issuing The Boys each a toothbrush, a tube of toothpaste, and a spool of dental floss, Dr. Armstrong further instructed Alonzo and Benjamin on the practice of daily, oral hygiene, and gave them a follow-up schedule when he would then remove and fill the cavities in their teeth.

Annual physical and oral examinations were promoted and encouraged by Mr. Smithfield by paying his employees while they attended to their health. Mr. Smithfield's business model ensured that a healthy employee was a profitable asset.

On November 30, 1919, Pastor Moses presented Alonzo and Benjamin each with their first paycheck, which they had earned for their labor and skills. Regardless of the day of the week, laborers were paid on the last day of the month. After deducting $1.50 for pay roll taxes, and $5.00 for rent, The Boys' net pay for the month of November was $38.50 each. Receiving cash payments for their skills and labor encouraged Alonzo and Benjamin to stay and work for the Smithfield Plantation.

Mr. Smithfield's formula for paying his employees encouraged loyalty and long-term retention. The compensation allotted to the laborers required for them to work five and a half days per week. This method of payment encouraged the laborers to attend worship services, and with their time off on Saturday afternoons, to help the less fortunate sharecroppers with their work chores.

Alonzo and Benjamin volunteered much of their labor and skills to assist sharecroppers with their field crops, butchering, and curing of their winter meat. For assisting the sharecroppers, The Boys were invited to Sunday dinners, and gifted with cured meat, canned fruit, and various types of cakes and pies.

Holding and looking at their pay checks in amazement, Pastor Moses seized the opportunity for a moment to teach the principles of basic arithmetic. "This is what each of your pay checks represent," began Pastor Moses. "For addition, let us use the plus sign. Each of you were paid for thirty days of labor at one dollar and fifty cents per day. Adding together $1.50 for each of those thirty days equals $45.00. For subtraction, let us use the minus sign. Each of you pay $5.00 per month to rent the hut where you live, plus $1.50 was withheld from your check to pay state pay roll taxes. Combined, that equals to a total of $6.50. Now, subtracting $6.50 from $45.00 equals how much?" Pastor Moses challenged The Boys' mathematic skills.

Using their fingers for calculators, Alonzo responded, "$38.50, sir!"

With Benjamin in agreement. "That is the right answer."

"That is correct. I am so proud for the both of you," Pastor Moses shouted, as he praised The Boys for paying attention while attending the adult class.

"So . . . now, what can we do with these checks?" a perplexed Benjamin inquired, while Alonzo acknowledged his brother's concern, with amazement on his face.

Continuing with the opportunity to teach, Pastor Moses instructed Alonzo and Benjamin to go with him to his home where he sat them at the dining room table. He went into his study room, and brought back four blank applications for opening a savings account at the Farmers Bank of Wyatt, two for each boy. One of the blank applications would be used to teach and practice how to complete an application with their signature.

The teaching technique that Pastor Moses employed was to use letters of the alphabet to sequence the spelling of The Boys' names: A-L-O-N-Z-O and B-E-N-J-M-I-N, then correctly transfer those letters of the alphabet to the application-block, which called for the entry of one's name. Moving from print to cursive, the process was repeated until The Boys were confident that they could complete their application with all requested pertinent information, and sign both the application and their check with a written signature.

After signing their checks and applications for opening a savings account at Farmers Bank of Wyatt, Alonzo requested that Mother Moses complete the transaction with the following: Take out enough cash so they could pay their tithe to the church, pay the Ice Man for the weekly delivery of ice, pay for the purchase of fresh fish from the Wallace family, pay for the purchase of an iron, so they could iron their jeans and shirt for the purpose to be neat and clean for worship, and purchase a razor with razor blades for each of them, so they could shave themselves of their facial hair.

Alonzo continued, saying, "We request that $5.00 be sent to Benjamin's mother, Ada, and James's mother, Flora, and inform Ms. Flora that her son, James, was deceased, and that we gave him a Christian burial in the state we believe to be Arkansas. Let them know both of us are well, and we are learning to read, write, and do arithmetic, and we will be sending a minimum of $5.00 a month to the each of you."

With tears in his eyes, and a tremor in his voice, Pastor took the hands of Alonzo and Benjamin, and while looking into their eyes, said, "I am overwhelmed with pride and thanks be to God. I am so blessed to witness the caring and sacrifices made by you, Alonzo, and you, Benjamin, and I am moved to promise you both that Mr. Smithfield and I will communicate with the owner of the Gilmore Plantation, and the spiritual leader for the negro population to make sure that your request be honored and successful."

The following week, on a Saturday afternoon, Mother Moses summoned The Boys to her home, and while sitting at the kitchen table, she prepared bologna sandwiches with the preferred condiments. Benjamin liked his sandwich with mayonnaise, while Alonzo preferred mustard, and, as a mother would know, Mother Moses gave each of her 'claimed boys', his favorite soda pop. She patted each on his head, and reported the good news to Alonzo and Benjamin on the requests they had made: (1). Reverend Luke Solomon, with his address, was the pastor for the negro Baptist Church of the Gilmore Plantation, and that James's mother, Ms. Flora had been informed of her son's death and burial. (2). Western Union Wire Services had been used to send $5.00 each to Ms. Flora, and Ms. Ada. And (3). After taking out of their checks the amount of cash The Boys requested, and purchasing groceries, Mother Moses presented Alonzo and Benjamin with their first savings account book, each showing a balance of $2 4.90, paying compounded interest at .05 percent annually. "Are there any questions?"

"No mam," replied Benjamin.

"And we thank you, so very much," concluded Alonzo.

Faithfully reporting for work every weekday, volunteering their time and skills to help the sharecroppers with their chores, attending their adult class of reading, writing, and arithmetic, and tithing and attending worship services every Sunday morning, the life of Alonzo and Benjamin now had perfect orientation, with the routine of a purpose-driven life that went from days to weeks, and weeks to months, and months to years.

Chapter Seven

Wyatt A Countrypolitan

ALONZO AND ROSE CONTINUE TO REMINISCE

O ctober of 1926, the anniversary month of when Alonzo and Benjamin emerged from the woods of the Smithfield Plantation. The first structure they saw was the Baptist Church of Mule Ridge. A place to worship, and a symbol of hope.

Hope and faith had transformed two boys, who were frightened, wearing ragged clothes, and worn-out shoes, hungry, infested with intestinal worms, and illiterate, into brave, healthy, educated, and responsible young men.

The first Saturday of October was the beginning for the week-long carnival hosted by the Countrypolitan of Wyatt, which was the center of commerce and finance for the surrounding towns, communities, and villages, which included, but was not limited to: Charleston, Brewers Lake, Byrd's Point, Deventer, Alfalfa Center, Pin Hook, and many others within a fifty-mile radius.

The population of Wyatt numbered 3,042, which would double on any given day during the week-long carnival.

Wyatt practiced "Jim Crow Laws, and Segregation," with highly visible signs for usage of certain facilities such as bathrooms, drinking fountains, seating in restaurants, and the movie houses, that read "WHITE ONLY" or "COLORED ONLY".

Railroad tracks for passenger and commercial trains divided and segregated the Countrypolitan Town of Wyatt into white and negro. The businesses that supported the financial base for the negro section of Wyatt consisted of restaurants, sandwich shops, and juke joints for entertainment, and serving alcohol.

The negro high school, which taught grades 9-12, was named Lincoln High School. It had a practice of cotton-chopping and cotton-picking vacations, which limited the negro students to an estimated school year of five months.

The businesses that were the commercial and financial engines that drove the white section of the Countrypolitan Town of Wyatt, included, but were not limited to: (1). Gladstone Fairweather Health, and Dental Clinic, with a seven-bed capacity for the treatment of patients. (2). Three cotton gins. (3). One lumber yard and saw mill. (4). One slaughter house and meat packing company. (5). One grain storage and silo company. (6). Once a day passenger and commercial train schedule. (7). Twice a week passenger bus schedule. (8). Farmers Bank and Savings of Wyatt. (9). Two gasoline and service stations. (10). One movie theater, with a seating capacity of fifty for white only, and twenty for color only movie goers. (11). One pharmaceutical drug store, and confectionary. (12) One liquor, wine, and beer store. (13). One dry-goods and clothing store, with a catalogue ordering department. (14). Three grocery stores, and (15). A United States Post Office.

The white high school was named The Wyatt High School, and was in session for nine months, and taught grades 9-12.

Since arriving at the Smithfield Plantation six years earlier, Alonzo and Benjamin had accomplished many milestones. Among them were securing full-time jobs, at $1.50 per day, to their current wages of $3.00 per day. They had now grown out of the fear of being captured and returned to the Gilmore Plantation, so they felt comfortable traveling into town without the supervision and protection of Pastor Moses or Mother Moses.

Transportation to the annual carnival, which was hosted by the Town of Wyatt, was provided and financed by the Smithfield family. The Mule Ridge Baptist Church was the designated point of departure and return for the negro population. The first of several departures began at 8:00 a.m. on the first Saturday of October, with the last bus scheduled to leave the carnival at seven in the evening. Alonzo and Benjamin, eager and excited to experience the joy and thrill of a carnival for the first time of their life, departed on the first scheduled bus, and returned on the last.

The Boys, who now were young men, were dressed as if they were twins, as they were quite similar in appearance. They wore matching brown western boots, hats, and starched, iron-creased blue jeans. The young men struck a handsome and confident profile.

Equipped with the abilities to read and write, and understand basic arithmetic, Alonzo and Benjamin had with them their Wyatt Farmers Bank and Savings account book. Using the for "COLORED ONLY" bank teller, the young men withdrew fifty dollars each from their respective accounts, which they used to pay for various rides offered by the carnival, hot dogs, hamburgers, candy bars, and various flavors of soft drinks.

The dearest, and most important purchase that Alonzo and Benjamin made on their first unsupervised trip to the city of Wyatt was a Christmas gift for Pastor and Mother Moses. On previous Thanksgivings, and Christmases, Alonzo and Benjamin had been the beneficiaries to receive gifts and blessings, but this holiday season they planned to say thanks with a gift.

The year of 1926 ended with a joyous, but rain-soaked holiday and Christmas celebration. Alonzo and Benjamin prepared Christmas dinner for Pastor and Mother Moses. The dinner consisted of chicken, with corn bread stuffing, collard greens, and a peach cobbler for dessert.

The gift, neatly wrapped in red and green wrapping paper, was carefully opened by Mother Moses. It was an oakwood stained plaque, measuring fourteen inches in diameter, with hand-engraved words created by Alonzo and Benjamin that read: "FOR GOD SO LOVED ALONZO AND BENJAMIN THAT HE GAVE TO THEM PASTOR AND MOTHER MOSES."

With prayers that were filled with pride, and emotional joy, as parents would have for their biological children who had demonstrated responsibility, care, and love for their mom and dad, Pastor and Mother Moses returned the same love and caring by presenting Alonzo and Benjamin with a new set of encyclopedias, and encouraged the young men to "always yearn for and seek knowledge." Mother Moses reminded the young men to wire a financial gift to Misses Ada and Flora, noting with a motherly tone in her voice that the gift should now be for $10, every month.

Chapter Eight

Flood and Death

Alonzo and Rose continue to Reminisce

The new year,1927, began where the previous year ended, with record setting rain fall throughout the Southern, and Mississippi Valley states. The record setting fall-season of rain, caused the Mississippi River to over flow its banks, and flood Mule Ridge, and the Smithfield Plantation.

In late March of 1927, the flood waters of the Mississippi River had engulfed and inundated the fenced and gated pastures, and much of the farm land of the Smithfield Plantation.

The natural topography of Mule Ridge consisted of small hills and ridges, thus its name's sake. Because of its mule breeding reputation, in conjunction with the natural landscape, the name Mule Ridge was a natural and fitting blessing during the flood.

As the flood waters slowly engulfed the Smithfield Plantation, the gates to the fenced pastures were opened, allowing the animals to follow their natural instincts for survival to move to the higher hills and ridges.

With the ending of the record-setting rainfall, combined with the advancing of the spring-season ushering in warmer weather, and the unmatched assistance provided by the United States Corps of Engineers, who provided mechanized equipment, an assortment of both motorized and conventional boats, tents for living quarters, and military personnel who provided assistance both day and night to transport food for the animals that were stranded on hills and ridges, the Smithfield Plantation of both human and animal life survived the flood of 1927. There was no loss of human life, and the loss of animal life was negligible, due primarily to the spring-season when animals were giving birth to their young.

In late May of 1927, the flood waters had completely receded back into the banks of the Mississippi River, and with its receding came the arduous task of locating and safely securing the animals, cleaning the barns and stables, and cleaning the homes of the residents.

With the loss of income from the mule breeding operation, the beef and pork operation, the logging and timber operation, and the agriculture crops, especially the cotton crops, along with expenditures to purchase food for the animals, took a physical, emotional, as well as a financial toll on the personal health of Mr. Smithfield.

On October 31, 1927, after a demanding day of physical work, and supervising the final stages of cleaning the barns and stables, Mr. Hunter Smithfield Sr. died while asleep, from an apparent heart attack. He was sixty-two.

Chapter Nine

Alonzo's Mind's Eye and Hope

ALONZO AND ROSE CONTINUE TO REMINISCE

O n the first Friday of November, 1927, at 11:00 a.m., a home going celebration was held for Mr. Hunter Smithfield Sr. at the white Baptist Church of Mule Ridge. Reverend Luke Solomon, the pastor, was the officiant.

The following day, Saturday, from the hours of 2:00 to 4:00 p .m. at the negro Mule Ridge Baptist Church, Mr. Smithfield's remains lay-in-state for a memorial service officiated by Reverend Aaron Moses, who opened the service with the reading of Psalm Number 23 (KJV), while acknowledging the presence of Mrs. Adaline Smithfield, widow of the deceased, and Ms. Rose Broussard.

After their half day of required work, many of the negro laborers, and sharecroppers, attended the memorial service, noting with sincere, and heartfelt respect, that "Mr. Smithfield was a good man, and that he sincerely loved and respected the negroes, and sharecroppers who lived and worked on his plantation." They all prayed for God to allow his soul to "Rest in Peace. Amen!"

After reciting the benediction, and reading chapter three of Ecclesiastes (KJV), the mourners began the procession to shake hands, and give their condolences to Mrs. Smithfield, and Ms. Rose.

With a warm, gentle, but strong handshake, the eyes of both Alonzo and Rose met in a tender and meaningful gaze that caused Alonzo's mind's eye to flash forward to what he dreamed the physical attributes of a wife for him to be. He pondered this moment in his heart.

Mr. Hunter Smithfield Sr. was buried in a grave along the side of his son, Hunter (Little Hunter) Smithfield Jr., in the white section of the Mule Ridge Cemetery.

Before his death, Mr. Smithfield had made arrangements for the families of laborers, and sharecroppers to receive a turkey and or meat of their choice for their holiday dinner; however, the Thanksgiving and Christmas holiday season were celebrated with a layer of grief and mourning for their benevolent benefactor, Mr. Hunter Smithfield Sr.

After exchanging Christmas gifts, while still seated at the dinner table, Alonzo requested that he and Pastor Moses be excused from the company of Benjamin and Mother Moses, as he needed personal counseling and advice from the pastor.

In the privacy of Pastor Moses's study, Alonzo shared with Pastor Moses his experience when he shook hands and gazed into the hazel-grey eyes of the young lady who sat on the front row pew along the side of Mrs. Smithfield.

"I have never met her, but a picture of that young lady has been in my mind's eye for several years, ever since I began to imagine what features I wished for a wife to have. Her complexion, similar to my mixed-race skin color, her long black wavy hair, thin nose, and hazel-grey colored eyes. I believe her to be about 5 feet 6 inches in height, and weighs approximately 120 pounds," concluded Alonzo, by describing the young lady.

"That young lady would be Rose Broussard. My wife, Angel, was the midwife who assisted with her birth and delivery. Rose will soon be turning fourteen years old, but she is very mature, both physically and mentally. Allegedly, she is the granddaughter of Mr. and Mrs. Smithfield," Pastor Moses volunteered, as he shared Rose's life history with Alonzo.

"Can I meet her?" Alonzo inquired, while rubbing his hands together, with his eyes focused downward toward his feet.

"I am sure you can," Pastor Moses exclaimed with a smile.

"Am I too old to court her?" Alonzo asked.

"Oh, I don't think so! I am twelve years older than Angelica," Pastor Moses said with a tone of encouragement in his voice, while instructing Alonzo the biblical story of Jacob and Rachel, as recorded in the 29th chapter of the Book of Genesis.

"Alonzo, have faith that what God has for you is for you, and that with his infallible wisdom, God is the one who allowed you to meet Rose. I will be speaking with Mrs. Smithfield for the purpose to schedule a day and time for you and Rose to formally meet," Pastor Moses promised Alonzo.

Pastor and Mother Moses said good night and wished the young men a Merry Christmas, and a Happy New Year!

Chapter Ten

Alonzo and Rose Meet

ALONZO AND ROSE CONTINUE TO REMINISCE

On January 31, 1928 at 1:00 p.m., Pastor Moses, along with Alonzo, as was customary for negroes to enter the home of a white person, knocked on the back door of Mrs. Adaline Smithfield. Mrs. Adaline opened the door. "Good afternoon," she said, with a bashful and smiling Rose by her side. "Alonzo, this is Rose, my grand-daughter. I believe you'll have met.

"Um, yes ma'am," replied Alonzo, while rubbing the side of his face.

"Well, you'll come on in," Mrs. Adaline said, with a smile and soft giggle, as she held the door open, with one hand, and placed the other on the shoulder of Alonzo.

"Thanks, for receiving us into your home," replied Pastor Moses, as he reached out his right hand, and gently shook the hand of his host.

"Yes, ma'am, thanks," joined Alonzo, with his right hand extended.

Seated comfortably on a white and gray, high-backed sofa, after an exchange of warm introductions, Rose offered beverages to Alonzo, who was obviously excited, and Pastor Moses, Alonzo's confidant and counselor, but her gracious hospitality was declined.

Mrs. Adaline and Rose each sat in high wingback chairs, which matched the sofa. The twelve by fifteen-foot living room was comfortably heated, with a three by fifteen-foot, open-hearth, wood burning fire place that was facing an eight by six-foot picture window. The ten-foot high, egg-shell white walls were decorated with Norman Rockwell, and New Orleans, Louisiana themed paintings and drawings. The rest of the furniture consisted of a small oak table with two matching chairs, and an upright oak-finished radio-combination gramophone.

Mrs. Adaline, breaking the atmosphere of quiet nervousness, made the announcement, "Alonzo! I have been informed that you wish to court my *precious, precious* granddaughter, Rose?"

"Nanna!" Rose exclaimed with a mixture of quietness, and stuttering in the tone of her voice, wanting her Nanna to stop speaking, but at the same time wanting Alonzo to know that she was extremely attracted to him.

"Alonzo, are those your intentions?" Mrs. Adaline continued, with a stern tone in her voice.

"Umm, umm . . . Yes ma'am," Alonzo finally responded.

Giggling softly, Pastor Moses gently encouraged Alonzo to confidently speak up, because he had shared with Mrs. Adaline the complete content, with context of their conversation describing his experience, while shaking hands with Rose, as they gazed into the eyes of each other.

"Well! Alonzo, for your information, Rose has shared with me that she, too, experienced the same out flow of emotions, and she has been pressing me constantly, since that hand shake and gaze that the two of you shared, to find a way for the two of you to formally meet," said Mrs. Adaline.

"Nanna! I intended for you to share that information with Pastor Moses, so that he could tell Alonzo that I was attracted to him!"

"Well, now it is done," Mrs. Adaline said, with a smile and an approving chuckle. "Alonzo, I understand that you have some concerns due to the age differences between you and Rose. You soon will be twenty-four, and Rose will turn fourteen on February 3rd, just three days from today. I am sure that is going to be a blessing. My loving and late husband, Hunter, was twelve years my senior. In all aspects of life, Hunter taught me what goes into making a relationship physically, emotionally, and sexually healthy. I was fifteen years old when Hunter and I were married, and I have not regretted a single day. I miss him so very much," confided a teary eyed Mrs. Adaline.

Before setting the rules and conditions for the courtship to begin between Alonzo and Rose, with her blessings, Mrs. Adaline congratulated Alonzo, along with his brother Benjamin, for their courage and strength to overcome the hardships and ordeals they were faced with upon running away from the Gilmore Plantation in Tupelo, Mississippi to the Smithfield Plantation.

"Alonzo, you are a strong young man. However, these are the rules that you and Rose must obey," Mrs. Adaline began. "First, your visits, fun dates, and outings will be under my supervision, and or the supervision of Pastor and Mother Moses. Second, these activities will take place on Saturday and Sunday afternoons, and will end at nine o'clock in the evening."

Before adjourning the meeting, Mrs. Adaline made one more note. "Oh, and Alonzo, before Hunter and I were married, I was a virgin, and Rose, likewise, is a virgin. And I expect her to remain so until she is married."

"Nanna! That is too much information," Rose admonished her Nanna. She blushed and momentarily avoided eye contact with Alonzo.

"I am also a virgin," exclaimed Alonzo, respectfully, and then he and Pastor Moses said goodbye, and returned to their homes, riding in Mother Moses's mule-drawn buggy.

Chapter Eleven

A Year to Never Be forgotten

ALONZO AND ROSE CONTINUE TO REMINISCE

B efore Rose's feet could safely reach the clutch, brake, and gas pedals, "Big Daddy," the name Rose used to privately address her grandfather, Mr. Hunter Smithfield Sr., would allow Rose to sit in his lap and steer his automobile while he managed the pedals, in the process of teaching his granddaughter the safe mechanics of driving a stick-shift vehicle. At the age of ten years old, Rose was an accomplished and safe driver of an automobile.

At approximately 2:00 p.m., on February 3, 1928, Rose arrived at the home of Pastor and Mother Moses, with her Nanna being the passenger, driving a black 1925 two-door Hudson sedan. Mrs. Adaline, Pastor, and Mother Moses, along with Benjamin and Alonzo had successfully planned a birthday party for Rose. It was her fourteenth birthday. The celebration was not only to honor Rose on her birthday, but for each person present, to acknowledge, and give their individual prayers and blessings for the courtship of Alonzo and Rose. It was an honored tradition.

They prepared to serve the honored guest her favorite meal, which included fried catfish, prepared and deep-fried by Benjamin, chicken and dumplings, cabbage, okra, and corn bread, with a chocolate-icing-topped devil's food cake for dessert.

During the celebration, and among the multitude of conversations that were filled with joy and laughter, Mrs. Adaline and Mother Moses began to reminisce about when Rose was born and Mother Moses was the attending midwife for Charlotte Rose Broussard. The two fondly remembered what a beautiful baby Rose was.

"And it seems like it was just yesterday!" noted Mrs. Adaline, and Mother Moses nodded her head.

"Now, she is a beautiful and stunning young woman. I don't blame Alonzo for wanting to get his hands on her," a smiling Mother Moses noted with a chuckle.

"Yup, you got that right, and she can't wait for him to get his hands and everything else on her," added Mrs. Adaline, as the two recalled the days of their youth.

Rose was presented with an assortment of gifts, which included skirts, blouses, shoes, and an assortment of under garments. But the most memorable moment was when Alonzo presented his gift to Rose, a gold, chain-linked bracelet with the individual links measuring in size the diameter of thin fishing-line. Hand shaking from nervousness, Alonzo summoned the help of his brother, Benjamin, for assistance in placing the bracelet on Rose's wrist, which was thin and shapely. The moment was filled with relief and laughter.

The joy-filled celebration concluded with warm embraces and goodbyes. Alonzo and Rose, holding hands, walked slowly and clumsily to her automobile. Alonzo, like a gentleman, opened the passenger door for Rose to enter.

With perfect timing to end a memorable occasion, Pastor Moses shouted while standing on his porch, "I am expecting for the both of you to be present for Sunday morning worship!" In doing so, he rescued Alonzo from a moment of not knowing what to do.

Obeying the command of Pastor Moses, Rose drove herself to the Sunday morning worship hour, joining Alonzo and Benjamin on the front row pew.

Absorbing the sermon and lesson taught by Pastor Moses, from the Book of Exodus of how God had delivered the children of Israel out of slavery from the land of Egypt, the three, Alonzo, Benjamin, and Rose felt touched because of how God had delivered the each of them from their personal experiences of hardships, and dependance on others. With the spiritual experience and associating their personal experience with the sermon and lesson, each, when the doors of the church were opened for accepting Jesus Christ as One's Lord and Savior, led by Alonzo, then Benjamin and Joined by Rose, proceeded to the Altar. They accepted the right hand of fellowship with all rights and privileges from Pastor Moses, and members of the congregation to be baptized into the Christian faith, and become members of the Mule Ridge Baptist Church.

"Baptism for these three, who have accepted Jesus Christ as their Lord and Savior, will be performed on the first Sunday of July, that is when the water in the Blue Hole will be warm enough for full immersion," announced Pastor Moses, while giving the benediction and prayer. Amen.

Thereafter, Rose, driving herself in the black Hudson sedan, would arrive an hour before the start of the Sunday worship hour so that she and Alonzo could spend more time together for talk, and to get to know each other more deeply. Furthermore, Rose presented a plan to her Nanna, where she would teach Alonzo how to drive an automobile. This plan allowed the two of them to hug and kiss, and experiment with "innocent" petting. With the opportunity for embraces and prolonged kissing, coupled with petting, Alonzo and Rose remained obedient to the demand of Mrs. Adaline, and maintained their virginity.

Utilizing every opportunity to be in the company of each other, Rose noted that Alonzo's official birthday would fall on the 29th of February, a leap year. Presenting this information to her Nanna, Rose proposed that an appropriate gift for Alonzo would be for the two of them to attend a movie together.

Accepting Rose's proposal, Mrs. Adaline exclaimed with laughter, "Rose! Honey child, you are absolutely getting the most out of my rules and conditions that the two of you could only be together on Saturday and Sunday afternoon, with the supervision of myself, or Pastor and Mother Moses." Rose's Nanna dropped off the two at the Wyatt's movie theater to attend the seven o'clock evening showing, and that she would arrive to pick them up at 9:00 p.m. sharp.

After being picked up in front of the movie theater, and on the way to being chauffeured to their respective homes, Rose and Alonzo, holding hands, sat in the back seat of the Hudson sedan when Rose began to share with her Nanna what she and Alonzo had experienced.

"Nanna," Rose began with hurt resonating in her voice, "we were instructed to sit in the 'COLORED ONLY' section of the theater. But when Big Daddy went to the movies with you and me, we never had to sit in that section!" Rose exclaimed, while rolling her eyes, and grabbing her head with both hands.

"Yea. That's just the way it is," Mrs. Adaline responded, while shaking her head.

"But anyway," Rose continued with a review of the movie which was titled *Woman of Affairs*, "the stars of the movie were John Gilbert and Greta Garbo, and the theme told the story of two childhood sweethearts who were kept from marrying, and misery ensues!" Rose noted on completing her review before they arrived at the home of Alonzo.

"Good night Mrs. Adaline. Good night, Rose. Thanks so very much for my birthday gift. It was the first time for me to attend a movie," Alonzo confided bashfully, and he again thanked Mrs. Adaline for chauffeuring he and Rose to the movie theater.

Retiring to her bedroom, and while preparing for bed, Mrs. Adaline pondered in her heart what Rose's next plan would be.

Spending every minute that was possible with Alonzo, and while studying for her final examination to satisfy the requirements for the State Board of Education for home schooling, time seemed to fly for both Alonzo and Rose.

After Alonzo learned how to drive an automobile from Rose, Mrs. Adaline gave Alonzo permission to use the same vehicle to teach Benjamin the skills to safely drive a vehicle.

During the month of June, 1928, Rose received official notification from the State Board of Education that she had met the requirements to satisfy a twelfth-grade certificate of graduation. Also, on the last Saturday of the month of June, in the afternoon, and off from work, both Alonzo and Benjamin successfully passed both the written and road tests for their driver's licenses.

The birth date for Benjamin, as it had been for Alonzo, had never been officially recorded. Benjamin believed that he was born two months later than his brother, Alonzo, and between the years of 1900-1908. Following after the reasoning that Alonzo had used, Benjamin chose the middle year, 1904. For a birth month and day, he

chose, July 4th, as it was fitting to memorialize the date that he, Alonzo, and James walked away from the Gilmore Plantation.

Benjamin's driver's license officially recorded him to be:

Benjamin Gilmore

Date of Birth: July 4, 1904

Race: Negro

Complexion: Light Brown Skin

Color of Eyes: Grey

Hair: Black, and Wavy

Weight: 180 pounds

Height: 6'-1"

Except for the first name and date of birth, Alonzo's driver's license
recorded the same information. Alonzo and Benjamin were often
mistaken to be twins.

Paying the sum of seventy-five cents each, Alonzo and Benjamin
received their driver's licenses from the State of Missouri to operate a
passenger vehicle. The license was to be renewed every two years.

On close examination of their driver's licenses, while back in the vehicle with Mrs. Adaline and Rose, who had driven, and accompanied
them to the driver's license bureau, Alonzo and Benjamin discovered,
after all these years, that they were in the state of Missouri.

"What! We are not in Illinois, the Land of Lincoln and Freedom?"
exclaimed Alonzo.

Vigorously shaking his head, and laughing loudly, Benjamin, finally
said, "What? We have been in Missouri all these years. How is it
possible for us not to know that we were in Missouri?"

"It is because we did not know how to read, and write, and everyone whom we have met—Pastor and Mother Moses, Mr. and Mrs.
Smithfield, and many others—have received us and treated us with
respect, and love. Also, God has been watching over us, and love can

be found in all of the states, not just Illinois," said Alonzo, with a tone of thankfulness in his voice.

"Amen! God's divine wisdom kept y'all from attempting to swim across the Mississippi river at the Byrd's Point location," began Mrs. Adaline. "I know that each of you are strong swimmers. However, if y'all would have attempted to swim across the river at the Byrd's Point location—and that is where the confluence of the Ohio and Mississippi rivers meet—the currents and turbulences at that point in the rivers would surely have caused both of you to drown. Alonzo and Benjamin, if God had not intervened and stopped y'all at the Smithfield Plantation, each of you would have drowned. We are just two to three miles from where you would surely have met your 'Waterloo'. I am going to drive us there, and let you see for yourselves how dangerous that effort to swim to the Land of Lincoln and Freedom would have been," Mrs. Adaline concluded with her lecture.

Witnessing and acknowledging what surely would have been their "waterloo," Alonzo and Benjamin secretly pondered in their hearts the following:

I would never have met Rose, Alonzo secretly pondered in his heart.

Dear God, thanks so much for your divine intervention, and please help me to reach the Land of Lincoln and Freedom, along with my mother, Ada, and Ms. Flora, the mother of James, Benjamin prayed secretly in his heart.

It was now Sunday, July 1, 1928, the scheduled date for the new born believers in Jesus Christ to be baptized into the Christian faith. The water in the Blue Hole was measured at a comfortable 82 degrees Fahrenheit, with an attending cloudless, high-blue sky, and a very hot outdoor temperature of 102 degrees.

With an estimated crowd of a hundred worshippers standing beneath the shade of the tree-lined bank of the Blue Hole, and wearing various types of hats and head coverings, Pastor Moses chose the Gospel of Mark, chapter 1, verses 10-12 to teach and preach the spiritual significance of full immersion of Christian baptism. Noting that when John the Baptist fully immersed Jesus Christ in the water of the Jordan River, the Holy Trinity was present with their approval, with the voice of God, the Father, speaking from heaven, and the Holy Spirit in the form of a dove, alighting on the shoulder of Jesus Christ, who had just been baptized.

As Pastor Moses entered the warm water of the Blue Hole, wearing his white baptismal robe, he announced the first names of those who would be baptized. "Alonzo, Benjamin, Rose, Sadie, Ida, and Willie. Not in my name, but in the power vested in me by God the Father, Jesus Christ the Son, and the Holy Spirit, which rest, rule, and abide in all believers, I, Pastor Aaron Moses, now baptize you," proclaimed Pastor Moses as he immersed each born again believer into the water of the Blue Hole.

After the baptismal ceremony, Pastor Moses led the congregation by walking the short distance to the sanctuary, while the congregation sang the hymn of "Take Me to The Water to Be Baptized". The worship hour concluded with Pastor Moses serving Holy Communion of Bread and Wine to all baptized believers who were present, and giving the benediction and prayer, and wishing everyone a safe and happy Fourth of July.

With their spirits still high in the aftermath of their Christian baptism, while under the trusted supervision of Pastor and Mother Moses for their Sunday afternoon courting session, Alonzo and Rose, while holding hands and walking along the bank of the Blue Hole, expressed their committed love for each other, and made plans to request permission from Mrs. Smithfield for their formal engagement, and marriage. Alonzo and Rose planned to make their request to

Rose's Nanna during the Fourth of July celebration, which would be hosted at the home of Mrs. Adaline.

With perseverance, and clandestine searching, Rose eventually located the wedding gown that her Nanna wore on her wedding to Mr. Hunter Smithfield, Rose's Big Daddy. Presenting the two-foot by three-foot cedar-wood luggage that preserved the treasured gown to her Nanna, and prepared to introduce their plans—Mrs. Adaline interrupted Alonzo and Rose, while opening the luggage, which had perfectly preserved the "Victorian, Edwardian Vintage, white laced gown, with white elbow length gloves," that she had worn on December 25, 1895, when she and Hunter Smithfield were married. Mrs. Adaline, was fifteen.

"Alonzo! Rose! You have my permission and blessings to marry each other!" Mrs. Adaline exclaimed, while simultaneously causing to become moot, the plans that Alonzo and Rose had authored and rehearsed to present to Mrs. Adaline, requesting her permission for them to be married.

"Thanks, so very much, Nanna. When Alonzo and I stand to accept our vows of marriage, I want to wear the gown that you wore on your wedding to Big Daddy," said a teary-eyed Rose, with her voice filled with a tone of happy emotions.

"Thanks, Mrs. Adaline for allowing me to Marry Rose," I will always cherish her love, and protect her with my life, said Alonzo, with nervous relief in the tone of his voice, as he was saved from performing the rehearsed speech to request permission for he and Rose to get married.

"Each of you are welcome. And Rose, absolutely, yes you can wear my wedding gown on your special day of holy matrimony. I, truly, count this day to be a blessing, and if I had not permitted y'all to get married, the two of you would have eloped, and gotten married anyway. That is what Hunter and I had planned to do if my parents had not given their permission and blessings for me to marry Hunter." Mrs. Adaline had a just-visible smirk.

"Nanna, I am so sorry. I know that you and Big Daddy had plans for me to attend college, but Nanna, I have been home schooled, but there were no other children whom I could identify and socialize and be friends with, because I am racially mixed. I am neither negro nor white!" exclaimed Rose with tears in her eyes.

"Rose, darling, and Alonzo, I understand, and I am saddened to know how difficult it has been for each of you and your experiences of being racially mixed, but God as my witness, I believe that it has been the divine intervention of our Lord and Savior Jesus Christ, that you two have been put together," Mrs. Adaline said as she broke the uncomfortable air of tension, with laughter.

"Thanks, so very much Nanna. You always say the right thing," Rose said with a beaming smile, and eyes opened wide.

"Yes, being racially mixed, and at my age, I had begun to think that I would never be blessed with a wife," confided Alonzo as he asked Mrs. Adaline for a hug.

Holding back her tears, Mrs. Adaline broke the moment of tension with the command: "Let's get the party started."

"Nanna, we plan to get married on February 3, 1929, my birthday. I will be fifteen, the same age *you* were when you and Big Daddy were married," Rose said.

"Same age, and the same size. The gown should require little, or no alteration. After Alonzo has gone home, you can try on the gown to make sure that it fits perfectly." Mrs. Adaline smiled and she excused herself to go into her bedroom.

She returned in a few minutes, holding up her hand. "This here is the engagement ring that my late husband, Mr. Hunter Smithfield

Sr., placed on my finger while bending on one knee and asking me to be his wife," Mrs. Adaline recalled with a tone of nostalgia in her voice. "Alonzo, I want for you to take this one-half carat diamond ring, which has two one-quarter blue sapphire begets, that is mounted on a silver band, and place it on Rose's left-hand ring finger, while bended on one knee, and ask her: 'Rose will you marry me?' Rose, you will reply: 'Yes, Alonzo, I will marry you!'"

Acting out the instructions that Mrs. Adaline offered, Alonzo knelt on to his right knee, and with a wide-open mouth, and toothy smile, asked, "Rose, will you marry me?"

"Yes, to the man who knew me in his 'mind's eye,' I will marry you, Alonzo Gilmore, and will always remain faithful," Rose promised with a quiet, and committed tone in her voice.

"Now the two of you are formally engaged to become husband and wife," exclaimed Mrs. Adaline, happily. "Furthermore, Rose, the home that your grandparents, Andrew and Mabel Broussard, lived in while raising your mother, Charlotte Rose, I will legally deed to Alonzo Gilmore, and Rose Gilmore, husband and wife, joint owners, upon the date that y'all are officially married. Also, not forgetting Benjamin, Alonzo, each of you will receive a pay increase from five dollars per day, to ten dollars. Alonzo, you will need an increase in wages to be able to support a wife, and hopefully lots of children," Mrs. Adaline concluded with pride, and she informed the newly engaged couple that she would be signing contracts with home remodeling contractors to install electricity, telephone lines, indoor plumbing, and toilets, showers, and a bath tub, and a propane heating and cooking system. "I want your home to have the conveniences for comfortable living," Mrs. Adaline proudly concluded.

Both in amazement, Alonzo and Rose, shouted in unison, "Wow! Thanks!"

Rose squealed and they both hugged Rose's Nanna.

Back at his home, Alonzo, excited and filled with happiness, in minute detail, shared with Benjamin the events surrounding his proposal of marriage to Rose.

"WOOOOOOOOOW!" responded Benjamin, while congratulating Alonzo. "Big brother, she is beautiful, and sexy."

"And very intelligent," a beaming, and proud husband-to-be added.

"Alonzo," inquired Benjamin with a tone of wry envy in his voice, "does she have a sister?"

"No, I don't think so, but we will keep praying that God will send to you the right companion." Alonzo patted his brother on the back.

With his head bowed, and clasping his face in his hands, with a serious tone in his voice, Benjamin asked, "Big brother, do you remember the day when we discovered that we were in the state of Missouri, and not in Illinois, the Land of Lincoln and Freedom? I pondered in my heart, and knew beyond any doubt that I would have to get to the state of Illinois before I could realize my dreams and hopes for personal freedom," Benjamin said, with tears in his eyes, while recalling the night they ran away from the Gilmore Plantation.

"Yes, I remember," Alonzo said. "I pondered how, that if we had attempted to swim across the Mississippi River at the Byrd's Point location, we would probably have drowned, and I would not have met Rose, my future wife," Alonzo confided with words of thanks to God.

"Alonzo, God has answered your prayers of dreams, hopes, and a future with a beautiful, sexy, and intelligent woman, with a home to move into when you and Rose are married. I am so happy for you, and I thank God for protecting and watching over us on our journey away from the Gilmore Plantation, but I must move on," said Benjamin, while embracing Alonzo, with a tone of deep sadness in his voice.

"I understand, so let us have a meeting with Pastor Moses to discuss our plans, and seek his advice," replied Alonzo, while reckoning with the reality that the love of brothers inevitably grows toward separation, while the love for a future wife, grows toward oneness, and being together.

"Furthermore, can we discuss with Pastor Moses of how we can assist my mother, and Ms. Flora, James's mother, on how to leave the Gilmore Plantation?" Benjamin said with a tone of questionable hope, and sadness.

"Yes, we will certainly discuss those plans," Alonzo said as he placed his right hand and arm around the shoulders of Benjamin.

On the second Sunday of July, 1928, following the worship hour, Alonzo and Benjamin met with Pastor and Mother Moses and presented their plan for Benjamin to move to the state of Illinois, preferably to the city of Chicago, and financially sponsor his mother, Ms. Ada, and James's mother, Ms. Flora, to leave the Gilmore Plantation, and relocate with Benjamin.

With words of encouragement and support, Pastor Moses said, "Alonzo, please accept my congratulations and prayers on your engagement to marry Rose. She is a wonderful young lady, and will be a blessing to you as your wife. God brought you together, and I am sure that he will bless your marriage." Then continuing with plans for Benjamin to move to Chicago, Pastor Moses promised, "I will be contacting Reverend Matthew Vaughan, Pastor of the Zion Baptist Church of South Chicago, Illinois, for advice and assistance on how to carry out Benjamin's plans. I will contact Pastor Vaughan with a Western Union Telegram tomorrow, Monday. And, by the first Sunday of August, I will let you know what will be required," before acknowledging that Pastor Vaughan was a man of God, and Faith, and would get things done.

After waiting patiently for two weeks, Alonzo and Benjamin met with Pastor Moses in his pastoral office, following the morning worship hour on the first Sunday of August, 1928.

Opening the meeting with good news, Pastor Moses reported to Alonzo and Benjamin that Pastor Vaughan secured a single family, two-bedroom home, with indoor hot and cold running water, with a bathroom that had a bath tub and shower. The home was equipped with electric power, and was heated with steam radiators controlled by a thermostat. The home was fully furnished. Rent for the home would be $105 per month. The renters would be responsible for their electrical and heating cost, which was estimated to be $45 per month, but the water bill would be paid by the owner, making the monthly cost approximately $150.

The bus fare for each passenger, departing from Tupelo, Mississippi, arriving at Chicago, Illinois bus terminal would cost $37.50. The bus trip would take approximately twenty-two hours, due to stops, and layovers in various cities. The ladies were instructed to pack some food to sustain them during the trip. Furthermore, while communicating with Pastor Moses, Pastor Vaughan arranged job interviews for Benjamin with a meat packing company and a steel manufacturing corporation, and for the ladies there were a large number of job opportunities for cooks, domestic workers, and house cleaners.

Pastor Moses concluded his report by emphasizing, "Pastor Vaughan had noted that the rental property would be available until October 3, 1928, and that a first and last month of rent, and electricity usage, totaling $300, must be paid as soon as possible."

With excited relief, Alonzo and Benjamin informed Pastor Moses that, between the two of them, they had a combined savings of $4000, and that he and Mother Moses had their permission to withdraw from their accounts whatever funds were needed to successfully complete their plans.

"Angelica and I, along with input and instructions from Pastor Vaughan, and with God's help, will manage to get done the request that y'all have made," Pastor Moses said while encouraging Alonzo and Benjamin to use their newly acquired reading, and writing skills to communicate with Misses Ada and Flora that blessings were coming to each of them.

After saying thanks and goodbye to Pastor Moses, while walking to their home, Alonzo and Benjamin, remembering James, their brother, began with solemn emotions to thank God for how he had blessed each of them from the times of being scared, ragged, and hungry, to now being healthy, confident, having a job, and being financially secure, and looking forward to a productive, and prosperous future.

Being occupied with their daily work schedule plus overseeing and managing the remodeling of Alonzo and Rose's future home, time for Alonzo and Benjamin seemed to fly.

With the help of Rose's Nanna, the wedding was meticulously planned, including the fitting and ordering of a three-piece, black tuxedo suit, a white dress-shirt with French cuffs that were accentuated with black onyx, gold mounted cuff links, and a white bowtie, for Alonzo.

Time for Benjamin was filled with his mind imagining his move to Chicago, and seeing his mother and Ms. Flora for the first time since he, Alonzo, and James fled from the Gilmore Plantation, which was approaching ten years past.

It was now October, 1928, the anniversary month when Alonzo and Benjamin arrived at the Mule Ridge Baptist Church, and faced with a dark and uncertain future. They were approximately fourteen years old when their journey began on July 5' 1918. After walking away from the Gilmore Plantation, The Boys, as they were referred to then, arrived at the Smithfield Plantation during the month of October,1920.

Eight years later, on a crisp Saturday morning in October, of 1928, as the sun was rising to fill a clear bright day with sun light and a high blue sky, Alonzo, with Benjamin holding in his right hand a one-way ticket to Chicago, Illinois, stood as brothers at the Wyatt, Missouri train station.

Benjamin and Alonzo, each with an arm around the other's waist, and never having been separated, knew that this departure was the beginning of a new life for the both of them.

"Benjamin," Alonzo called out with a whisper of a voice, as he was desperately trying to keep from crying, "do you remember when we, along with James, ran away? All we had was in a burlap bag. We were scared. Just trying to get to the 'Land of Lincoln and Freedom'."

"Oh, yeah, but now I have some real luggage, and this time, I know where I am going, and I am not afraid of being captured by the KKK," Benjamin said with a tone of sad melancholy in his voice. He carried a small, cardboard, single piece of luggage.

"Don't forget to wear the sign that we made, which spells out your name, 'BENJAMIN', around your neck," Alonzo reminded his brother, so Pastor Vaughan would be able to recognize him when he arrived at the Chicago train station.

Choking back tears, and with a quiet gaspiness in his voice, Benjamin said, "This is not a goodbye, but a farewell. We will see each other again, and soon." He boarded the train.

"Farewell!" Alonzo called, as he choked backed tears while remembering the farewell he and Benjamin had said to James as they buried him somewhere in Arkansas!

As Alonzo watched the train fade out of his sight, tears welled up in his eyes. He walked slowly to Mrs. Adaline's automobile, which she had allowed him to use to drive Benjamin to the train station. The young men, having never been apart, requested this moment to be alone with each other to experience their separation, in privacy.

After an eighteen-hour train ride, at approximately 1:00 p.m. on Sunday, Benjamin, with the sign hanging around his neck, was met by Pastor Vaughan and his wife, Ruth, along with Deacon Mark Crenshaw, Chairman of the Deacon Board of the Zion Baptist Church, of South Chicago, Illinois.

After warm greetings and introductions, Deacon Crenshaw, being the chauffeur of a 1925, black, two-door, four-passenger sedan, drove Benjamin, who sat in the front seat, while Pastor Vaughan and his wife rode in the back seat, to his new home. The thirteen-hundred-square-foot, single family, white-painted house sat on a fifth-of-an-acre lot, with a manicured front lawn and a comfortable sized fenced-in back yard. It was located at 403 South Elm Street, Chicago, Illinois.

"Allow me. I will take your luggage inside. Here, take the keys, to your new home, and see if you like it or not," Deacon Crenshaw offered.

"It is nice, very nice. I like the furniture, and this couch is very comfortable," Benjamin said as he rose from the couch to go and examine the furnished bed room and bath room and shower.

"Benjamin, look here in the kitchen. We ladies of the welcoming committee have stocked your food pantry with a variety of canned goods, along with some corn flakes and grits, and in the refrigerator, there are eggs, milk, and fresh meat, along with bologna, and head cheese. And a loaf of bread is on your dining room table," Mrs. Ruth noted with pride. She placed her hand on top of the gas-cook stove, and indoor sink, as she turned on the cold and hot running water. Opening the cabinets, which contained an assortment of plates, saucers, cups, silver ware, and several sizes of pots, pans, and skillets, she added that Misses Ada and Flora would enjoy their kitchen, while motioning for Benjamin to inspect the bedrooms, which had new mattresses, a quilt, and new linen for each bed.

Pastor Vaughan reminded Benjamin that the church secretary would be sending Pastor Moses a Western Union Telegram to inform him and Alonzo that Benjamin had arrived, and was safe in his new home. "Young man, I want you to be prepared, and ready for your 10:00 a.m. job interview with the meat packing company, which is scheduled for tomorrow, Monday," Pastor Vaughan noted with a quiet, and assuring tone in his voice.

"How will I get there?" Benjamin asked with a curious tone of anxiety.

"Deacon Crenshaw will be driving you there. Be prepared to leave your house at nine o'clock tomorrow morning. Wear clean, and neat clothes, and make a good first impression," advised Pastor Vaughan, noting, "That is where Deacon Crenshaw is employed, and he is the person who recommended you for the job."

The welcoming troop entered into the vehicle, and said goodbye, with Benjamin waving and pondering in his heart, *I have, finally, arrived at the 'Land of Lincoln and Freedom'!*

Chapter Twelve

Reunion, Marriage, Children, and a Mule Named Blue

The months of November and December of 1928 were filled with memorable events in the life of Alonzo.

Benjamin was now living in Chicago, Illinois, with full-time employment with the Swift Meat Packing Company. His forty-hour work week paid $3.00 per hour, and $4.50 for every hour of over time.

Ms. Ada and Ms. Flora had arrived safely to Chicago during the week of Thanksgiving, and with a happy and glorious celebration, were living with Benjamin in his two-bedroom rental home. The ladies were now employed at a reputable restaurant, as cooks.

Thanksgiving celebration for Alonzo and Pastor and Mother Moses, with Rose as a special guest, was celebrated with quiet, and solemn thoughts and prayers for Benjamin, and the bright future that was before him.

After their traditional Thanksgiving dinner at the home of Pastor and Mother Moses, Alonzo and Rose drove to their future home to inspect the progress of the remodeling. While at the home, and seated on the side of the American Iron Bed that Rose slept in while growing up under the care of her grandparents, Andrew and Mabel Broussard, Alonzo and Rose began to embrace, kiss, and pet each other passionately, like never before.

Rose, wearing a loose-fitting white blouse, and multi-colored dress, and thigh-high nylon stockings, took Alonzo's right hand, and placed it between her legs, and inside of her now moist panties. Moving their bodies farther onto the bed, while facing each other, Rose slowly unbuttoned her blouse, revealing to Alonzo her ample breasts, and hardened, cherry-sized nipples, which were supported with a sheer-white nylon brassiere.

Rose whispered, "I'm so wet! I have never been so wet, even when reading romance novels!"

"I know! I can smell the woman that is emanating from between your thighs. Rose, baby, I'm so hard. I'm . . . I'm about to explode."

"I know—Alonzo, you are so big! Your rod has to be ten to twelve inches long. You . . . you can make love to me, but please be gentle. *It is so big!*"

"Rose, baby, no. We promised your Nanna that we would remain virgins until we were married. We must keep the promise that we made to Mrs. Adaline."

"Nanna would never know," Rose pouted with frustration in the tone of her voice, while removing her wet panties, and putting on dry underwear, which were in the linen drawer of her bedroom dresser.

After calming their sexual arousal, and correcting their rumpled undress, Rose, while driving, stated, "Oh, Alonzo, I forgot to tell you. Nanna has made train reservations for the both of us to travel to New Orleans, Louisiana, in hopes of locating my mother, Charlotte Rose. Nanna, regardless of the lack of the success the private investigators have had, she insists on trying to find my mother, by herself."

Mrs. Adaline wished for Charlotte Rose to be present for Rose's marriage to Alonzo.

"Nanna and I are scheduled to depart for New Orleans on the morning of the first Friday of next month, December, and return during the evening hours of December 22nd."

"Nanna wishes for you to drive us to the train station. Will you"?

"I will, but only if you promise to return to me."

"I promise to return, and I will purchase something that is sexy and naughty for our wedding," Rose said with naughtiness in a throaty tone of voice. She and Alonzo said goodbye, after a long and passionate embrace filled with opened-mouth kisses.

To assist with locating Charlotte Rose Broussard, who now would be approaching age thirty, and having a photograph of her when she was approximately fifteen years old, Mrs. Adaline used the full weight of her maiden name, which was Boatwright. While Mrs. Adaline had relatives, both biological, and the Smithfield in-laws, living in New Orleans, she yet chose for Rose and herself to stay in a hotel. Being the protective grandmother, Rose's Nanna feared that her mixed-race granddaughter would not have been received with respect.

The Boatwright family, who were multi-millionaires, amassed their enormous fortune from the ownership of banks throughout the state of Louisiana. In addition to using the weight of her family's reputation and wealth, Mrs. Adaline also used her power and prestige of her marriage into the Smithfield family, who had accumulated their wealth and political influence from the ownership of millions of acres of farmland and plantations throughout the state of Louisiana.

However, with the use of a rental car, and driving fifty to one hundred miles per day, and visiting numerous Creole villages and hamlets during a ten-to-twelve-hour day of searching, calling in favors from politicians, business associates, family, and friends of the Creole communities, no one had seen or heard of Charlotte Rose visiting or living in New Orleans.

On their twenty-one-hour train ride to their home of Wyatt, Missouri, Mrs. Adaline and Rose accepted the sad and dark reality that Charlotte Rose may never be found.

With hope burning eternal, Mrs. Adaline kept in her employ, and under contract for the private investigators to continue their search for Charlotte Rose.

Breaking, and moving on from the cloud of sadness and defeat, to the brightness of hope, joy, and mental and emotional victory, Mrs. Adaline assured Rose that with the "sweet and naughty" lingerie she had purchased at the lingerie shop, while shopping in New Orleans, would certainly be a winner on the night of her wedding.

The Christmas of 1928, would be the last for Alonzo to spend as a lonely bachelor. His list to shop for Christmas gifts, that was provided to him by Rose, his soon to be wife, consisted of functional gifts. Rose's shopping list requested for Santa Claus to fill her basket with a journal, and an album for photographs, which she would use to record with words and photographs the life story of the many children that God would bless her and her husband with.

The gifts that Rose requested for Santa Claus to fill Alonzo's Christmas basket with, included lots of warm and comfortable work socks and shoes, knowing that her future husband's work required for him to be exposed to wet and cold environmental elements. Rose believed that these work conditions would be cause for Alonzo to contract some type of disease.

After enjoying a Merry Christmas and Happy New Year with Pastor and Mother Moses and Mrs. Adaline, January of 1929, for Alonzo and Rose was filled with the routine of their work, attending Sunday's worship, and church services. In addition to these routines, were twice-a-week wedding rehearsals, that were demanded and managed by Mrs. Adaline, who was a perfectionist.

Successfully applying for and receiving a marriage license, at the cost of $5.35, and completing blood tests at the Gladstone Fairweather Clinic for the cost of $7.25 for both groom and bride. Blood tests, before a couple could legally be married was required by the State of Missouri Health Department to prevent the spread of contagious and hereditary diseases. The results of their blood tests allowed Alonzo and Rose to legally become husband and wife. The self-discipline of abstinence would soon be over. Alonzo and Rose were still virgins.

On Rose's fifteenth birthday, February 3, 1929—a bright, clear blue, sun-light-filled sky, but a seasonally cold Sunday after noon—at three o'clock, Alonzo and Rose stood before Pastor Aaron Moses, who was wearing a black ceremonial robe, and in the living room of their future home, to take their vows of holy matrimony.

Rose stood wearing the gown that her Nanna wore on her special day, which was a white Edwardian, floor length gown, with elbow high white gloves, with matching white, three-inch high-heeled slippers.

Alonzo dressed in a black, three-piece tuxedo, with a white French cuff shirt, accentuated with black onyx-stone cuff links, and white bowtie, and wearing black Florsheim shoes.

The bride and groom presented a presence of royalty.

In attendance to witness the sacred marriage ceremony, were Mother Angelica Moses, Mr. Reginald Brown, professional photographer, and Mrs. Adaline Smithfield, who gave to Alonzo her granddaughter, Rose, to honor and protect in marriage.

Every one present was praying and hoping that a miracle would allow Charlotte Rose, the bride's mother, to somehow appear.

She did not.

After a prayer of faith and hope that God would bless the marriage between Alonzo and Rose to be fruitful, and filled with His Grace and Mercy, Pastor Moses gave the following instructions to the groom and bride:

"Alonzo Gilmore, and Rose Broussard, after I ask the following questions, each of you are to respond with the affirmative 'I do.' "Alonzo Gilmore, do you take Rose Broussard, in health and in sickness, in wealth and in poverty, until death, to be your lawful wife?"

"I do."

"Rose Broussard, do you take Alonzo Gilmore, in health and in sickness, in wealth and in poverty, until death, to be your lawful husband?"

"I do!"

Pastor Moses, now read from the Holy Bible, Mark Chapter 10, verses 5-9:

"[5] And Jesus answered and said unto them, For the hardness of your heart he wrote you this precept. [6] But from the beginning of the creation God made them male and female. [7] For this cause shall a man leave his father and mother, and cleave to his wife; [8] And they twain shall be one

flesh: so, then they are no more twain, but one flesh. ⁹ What therefore God hath joined together, let not man put asunder". (KJV)

Pastor Moses continued, *"Alonzo, you may now place the gold-band wedding ring, that you hold in your right hand, on the left ring finger of Rose."*

Alonzo slipped the ring onto Rose's finger.

"Rose, you may now place the gold-band wedding ring, that you hold in your right hand, on the left ring finger of Alonzo."

Rose slid Alonzo's wedding ring onto his finger.

"Now, with the power vested in me by Almighty God, I, Reverend Aaron Moses, pronounce Alonzo Gilmore and Rose Broussard husband and wife. Alonzo, you may now kiss your bride, Rose!" Pastor Moses exclaimed. *"God frowned on divorces, except in case of adultery. Amen."*

Anxiously looking forward to sharing their bed, and the consummation of their marriage, the newlyweds, with nervous impatience, accepted and thanked the attendees for their gifts and kindness.

From Benjamin was a card of well wishes, containing a US $100 bill, with instructions for the newlyweds to keep in case of hard times. Benjamin also asked his brother, and now sister-in-law, to forgive him for not attending their special day, but due to his recent employment, he did not have any accrued vacation.

Pastor and Mother Moses presented the newlyweds with a full set of pots and pans for their kitchen, and instructed the husband and wife to cook and eat healthy, home-cooked meals.

The photographer, Reginald Brown, blessed the couple with a gift card to have their first-born child recorded in a photograph with Mom and Dad.

As promised, Mrs. Adaline, Rose's Nanna, presented the husband and wife the deed to their home, plus the keys, and title to the 1925

Hudson automobile, which they had been driving. "I have purchased a new automobile," Mrs. Adaline said with a quiet giggle.

Ending with hugs and kisses, Pastor Moses suggested, "Let us all say goodbye, and allow Mr. and Mrs. Gilmore to begin their honeymoon!"

Rushing to their bedroom, Alonzo and Rose rushed to undress themselves, and each other. Rose excused herself to enter the bathroom.

Now wearing a revealing, and sexually appealing black negligee, Rose entered the bedroom to be greeted by her sexually aroused husband.

Rose, now his wife, was the woman Alonzo had visualized and fantasized in his mind's eye, who he wished a sexual partner and wife-to-be. Long black-wavy hair, framing a diametrically balanced face, with grey eyes, perfectly aligned pearl-white teeth, and chiseled chin, and a thin nose.

With her long-black hair flowing freely, and enticingly covering her ample breast, Rose allowed Alonzo to peek, and thrust for Rose's cherry-sized nipples that were now hardened and erect. Standing, and naked, Alonzo slowly pulled Rose into his arms as their hungry mouths met with passion. With a throbbing erection between Rose's sculptured thighs, Alonzo could feel the wetness of his wife's vagina, when Rose guided her husband's hungry mouth to her hardened nipples where he gulped them, while gripping Rose's buttocks, and sitting her on a towel placed over silk sheets that Rose had purchased for their bed, for their wedding night. Rose had anticipated, that still a virgin, she wanted to keep from staining the bed linen when her husband would penetrate her vagina, and tear her hymen.

Slowly positioning Rose's athletic legs, and thin, pedicured feet over his shoulders, Rose's suspicion, that her husband was blessed with a ten-to-twelve-inch penis, was true.

"Alonzo, honey, be gentle, and slow with me. Don't give it all to me at once," Rose pleaded with a quiet nervousness in her voice.

"Okay, baby, I will be gentle, and go slow. Let me know when to give it all to you," Alonzo assured, and comforted Rose, as his throbbing penis entered his wife's wet vagina and tore the hymen.

With Rose's long legs draped over her husband's shoulders, and his hands gripping her rounded and firm butt, Alonzo knew, and claimed his wife's virginity.

"Aaaagh!" Gasping with a faint scream, Rose felt the penetration into her vagina, and the tear of her hymen, and within a few minutes an outburst of pint-up joy and elation as she experienced her first orgasm, and begged for her husband to give it all to her.

"Oooh! Woow!" Alonzo groaned with excited joy, experiencing his first time to ejaculate, and at the exact moment that his wife reached her first orgasm.

For the remainder of the day light, and night hours, Alonzo and Rose made love, insatiably, while laughingly acknowledging that it was a blessing for them to have a week of vacation from work.

For the duration of their marriage, and seven children, Alonzo and Rose's love making was always in harmony.

The honeymoon for Alonzo and Rose lasted from their wedding day until March of 1931.

In the intervening weeks, and months, whenever the opportunity presented itself, Alonzo and Rose would attend theaters, drive-in-movies, or drive to neighboring states and cities, which included Cairo, Illinois; Paducah, Kentucky; and St. Louis and Cape Girardeau, Missouri.

The energy-filled and happily married couple maintained their work schedule and commitments made to the Mule Ridge Baptist Church, and volunteered to assist Pastor Moses with the teaching of the adult reading, writing, and arithmetic classes.

Also, during this period of time, Pastor Moses entered Alonzo into the ministry of deacon training, to become an ordained deacon of the Mule Ridge Baptist Church.

During the first week of March, 1931, Rose happily informed Alonzo that she had missed two consecutive months of her menstrual cycle, and that she was sure that they were soon to be parents.

"I had begun to think that I could not have children!" Rose confided, as she and the father-to-be embraced each other with long and loving hugs, while thanking God between sweet, passionate kisses.

Two weeks later, during the season when mares were giving birth to their mule foal, a first-time mother-to-be mare had difficulty while trying to deliver her foal. The attending veterinarian was fortunate to save the new born mule, but due to having to perform an emergency caesarian section to save the foal, the mare had to be euthanized.

"This tragedy occurred due to a mating and breeding error. The male donkey, because of the genetics of its breed, was physically too large to mate with the small breed of mare. This mating produced an oversized fetus, which the uterus of the mare could not safely give natural birth to," noted the licensed veterinarian in his official report.

Not being able to safely place the baby mule with a nursing mare, Alonzo took the mule to his home, where he and Rose successfully bottle fed and raised the animal.

Being pregnant with her first child, and her maternal instincts at a high, Rose took the lead and responsibilities in feeding and training the animal.

The entire color of the young mule's coat of hair was coal-black, and depending on how the sunlight would shine on its coat, the color of the mule would sometimes appear to be blue. Thus, Rose named the mule "Blue".

During Rose's pregnancy, with apples as a reward, Rose trained Blue to respond to a number of verbal commands, which included: "Come to me; follow me; kneel down; and come to the porch," which were used to allow the animal to be safely mounted. Also, Rose trained and conditioned Blue to comfortably accept a bridle and bits, and a riding saddle.

During the pregnancy, Alonzo would lay his head on Rose's growing abdomen, where the father- and mother-to-be would spend hours talking and reading aloud to the growing baby that lived in the womb of Rose.

Rose, careful and mindful of her physical conditioning and health, chose, when weather permitted, to walk the one-mile distance to and from work, instead of driving. As for Rose's work responsibilities, her Nanna, Mrs. Adaline, grew very dependent on her granddaughter's skills of managing the everyday business requirements of the Smithfield Plantation.

Continuing with the nurturing and training of Blue, Rose would reward the growing and eager-to-learn animal with an apple before leaving for work, and about one-half mile from home on her return from work, Rose would call for Blue to come to her, and reward the animal with another apple. This training technique allowed Rose to reinforce her training commands.

On August 10, 1931, Rose gave birth to her first child, a girl, whom she and Alonzo named Mary.

In attendance for the delivery, and happy occasion were, Mother Angelica Moses, the attending midwife, Alonzo, the proud father and husband, and Adaline Smithfield, Rose's grandmother.

Outside of Rose's bedroom window, where she gave birth to all of her children, stood Blue, as she instructed her husband to give to Blue an apple.

For the delivery of her next six children, the cast of witnesses would be the same, with Blue standing outside of Rose's bedroom window, where he would receive an apple.

On February 21, 1933, Rose gave birth to her first son, named Jonathan.

On January 9, 1935, Rose gave birth to her second daughter, Martha.

With the birth of Martha, Mrs. Adaline hired a building contractor to construct a new bedroom and bathroom to the home of Alonzo and Rose, so that the growing family would be comfortable. The girls would have their bedroom and bathroom, and the boys would have a separate bedroom, with a private bathroom.

On March 12, 1937, Rose gave birth to her second son, Jacob.

On February 12, 1939, Rose gave birth to her third daughter, Sarah.

On July 26, 1941, Rose gave birth to her third son, John David.

The interval of time for Rose's pregnancies, and deliveries, for her and her husband's children, were approximately eighteen months to two years.

Rose would not experience another pregnancy until late 1947.

Due to the experiences that Alonzo and Rose had endured growing up without a mother, and or father, each had vowed to never leave their children with anyone, including a temporary baby sitter. Therefore, to ensure that their children were taught values and principles that their parents espoused and adhered to, Alonzo and Rose scheduled their time at work for one of the parents to be with their children at all times. With her husband having to work during the daylight hours, Rose chose to work evenings and at night.

Alonzo continued to progress, and promotions added responsibilities with his employment at the Smithfield Plantation. As farmers became less dependent on mules to successfully till the land, plant the field crops, and transport their product to market, Alonzo was given the responsibility to transition the mule breeding operation to a mechanized operation, powered by gasoline-engines and farm equipment.

Just as he had progressed with his employment, Alonzo was also growing in his faith and responsibilities at the Mule Ridge Baptist Church, and on the first Sunday of October, 1935, Pastor Moses ordained Alonzo Gilmore to become a Deacon of the Mule Ridge Baptist Church.

Upon finishing with their meal, revisiting their years together, and reminiscing about the entirety of their life, Alonzo and Rose were prepared to share their life's history with their children.

Before the arrival of their children from their day at school, and placing the picnic meal in the wagon, which Blue had been hitched to, Rose made her husband promise that they would not share the details of their courtship and wedding night with the kids. Agreeing with adult giggling, the proud couple said, "I solemnly swear to never reveal what went on!"

Under the ripening nuts of a huge and glorious pecan tree, in the midst of the nut grove, and while enjoying their picnic meal, the children of Alonzo and Rose tentatively listened to their parents' life history.

Hearing their mom and dad share their challenges and ordeals of life's journey, Mary said, "I knew that we were different, with our complexion, color of our eyes, and hair, but we would never have guessed that we were that different." Jonathan and the younger siblings joined in with an outburst of loud laughter and clapping of hands, with their mom, noting that it was getting late in the evening, and they would have to gather nuts on another day, as the family proceeded to return to their home, with Blue pulling the wagon.

Chapter Thirteen

The Death of Rose

It was a chilly Thursday in April of 1948. Rose was now very late in her pregnancy with her seventh child. With the help of her husband, Alonzo, the proud parents had managed to serve their five school-aged children breakfast and kiss them goodbye, with each child carrying their lunch in a brown paper-bag.

After kissing, and seeing her husband off to work, Rose was left at home with John David, her youngest son. In the process of washing dishes, and cleaning the kitchen, Rose felt a sharp pain in her lower abdomen, and noticed that her under garments were wet. Upon further inspection of her underwear, Rose discovered that the wetness was caused from vaginal bleeding.

Not having experienced this symptom with her previous pregnancies, Rose took immediate action by preparing John David to go to where his father was at work, and inform his father that she needed him, and that it was an emergency.

Next, Rose called out for Blue to come to her, where she commanded the mule to kneel down, near the porch, to allow John David to use the height of the porch to mount the animal. Rose had trained Blue to perform this maneuver when he and her first child, Mary, were very young. It was Rose's practice to introduce Blue to each of her children when they were approximately two-weeks old. Blue had become a pet to the children, as well as their protector and guardian.

John David was now mounted and prepared, as he had done on previous occasions, to ride Blue bare-backed by supporting himself while holding onto the animal's mane.

Rose gave Blue the command, "Find Dad." One of the many commands she had trained the mule to obey. Rose then rewarded the animal with an apple, as Blue began a slow trot to find John David's father.

Being aware of the late stage of his wife's pregnancy, Alonzo was always alert, while at work, that he could be summoned to return home at any time. With this possibility constantly on his mind, Alonzo saw his son and Blue approximately one-half mile away from where he was working to prepare farm equipment for the upcoming cotton planting season.

Sensing an emergency, Alonzo hurried to his automobile, and sped away to meet his son and Blue. Helping his son to dismount the mule, and at the same time being informed by John David on what happened, the father, moving swiftly, while asking God to keep him calm and under control, opened the passenger side door. "John, son, hurry and get into the car," he said, and sped away to his home to give attention to his wife.

In the meantime, as he had been trained, Blue returned home.

Fighting the pain with the gnashing of her teeth and through pursed lips, Rose managed in a raspy tone to call out to her husband, before he entered into her bed room, "Alonzo, honey, go and bring Mother Moses, the midwife, and my Nanna. I am in labor," she said, while instructing her husband, through the pain of child birth.

When Mother Moses and Mrs. Adaline arrived, Rose, due to the loss of a large amount of blood, was in physical distress and experiencing severe pain. Mother Moses, utilizing her skills as a midwife, comforted Rose, and administered lifesaving support to the best of her abilities and training.

In the meantime, before leaving her home, Mrs. Adaline placed an emergency telephone call to the office of Dr. Fairweather, requesting that he proceed, with urgency, to the home of Rose Gilmore.

Dr. Fairweather was not available, as he was out of his office attending to in-home patients; however, Registered Nurse Florence, his daughter, would come to assist, and in the meantime, Nurse Florence made a telephone call to the Chief of Police of Wyatt, Missouri, requesting the help of his department in helping to locate her father, Dr. Fairweather, and appraise him of the emergency.

At approximately 2:30 p.m., after Rose had been in labor for more than four hours, Dr. Fairweather entered into the bedroom of Rose, which his daughter and Mother Moses, with lots of hot water, and clean towels, had converted into a make shift operating room.

With the professional nursing assistance of his daughter, and Mother Moses's midwife skills, Dr. Fairweather successfully performed a caesarean surgical operation to deliver Rose's baby girl.

The Gilmore children had arrived to their home from a day at school, and were huddled together in quiet anxiety in the family's living room. An exhausted, and weakened mother insisted that her children be allowed to enter into her bedroom, which now resembled a hospital emergency room. Blue, as always when Rose gave birth to her children, stood outside her bedroom window.

Having the entire family present, with the new born girl resting on her breast, and including her grandmother, Mrs. Adaline, Rose made several stern requests. First, the baby girl would be named Rachel. Second, each child would pursue their individual dream, and obey their heart. Third, each child must support each other. And finally, when it was time for Blue to go to mule heaven, that he was to be given a burial, and not discarded to a refuse scavenger service company.

With heavy hearts, and through tears of fear, and grief, each child vowed to honor their mother's requests.

After their children had reluctantly left their mother's bedside, Alonzo and Rose, while holding hands and reminiscing that they had done their best, and recalling when Pastor Moses officiated their holy matrimony, they had kept the vows they made.

"Until death do us part!"

To help control the pain, Dr. Fairweather had administered to Rose a high dose of pain medication, and due to the large amount of blood that she lost, while holding her husband's right hand, Rose slipped into a deep sleep.

She did not awake.

On April 22, 1948, just as the sun was setting, Dr. Fairweather, who was still present and comforting the children, pronounced Rose to be deceased.

She was thirty-four.

Chapter Fourteen

The Lessons That Life and My Father Taught Me

(THE MIDDLE YEARS)

The first Saturday of May 1948 was a perfect spring day. With a gentle southern breeze, the temperature was a pleasant 80 degrees Fahrenheit, with a very low level of humidity. There were no clouds in a high blue sky.

It was the day for Rose's home going celebration. She left to mourn, seven children. Four girls, and three boys, and her husband, Alonzo, of nineteen years. Rose Gilmore was thirty-four.

Reverend Aaron Moses, Pastor of the Mule Ridge Baptist Church, was the officiant.

After greeting, and welcoming the standing room attendees of mourners of approximately seventy-five, Pastor Moses acknowledged the family who were seated on the front row pew. Recognizing the family in order of relationship, order of birth, and age, were: Alonzo (husband), and holding Rachel, the newborn baby girl, Mrs. Adaline Smithfield (grandmother), Mary (oldest daughter), Jonathan (oldest son), Martha, Jacob, Sarah, and John David.

After a prayer for God to strengthen and comfort the family, Pastor Moses encouraged the family, and all mourners, to keep their eyes on the sky, and that while their spirit was now dark, God would allow the sunshine of joy and hope to shine in their soul, again.

After reading Chapter 13 of First Corinthian from the Holy Bible: *¹Though I speak with the tongues of men and of angels, and have not charity, I am become as sounding brass, or a tinkling cymbal. ²And though I have the gift of prophecy, and understand all mysteries, and all knowledge; and though I have all faith, so that I could remove mountains, and have not charity, I am nothing. ³And though I bestow all my goods to feed the poor, and though I give my body to be burned, and have not charity, it profiteth me nothing. ⁴Charity suffereth long, and is kind; charity envieth not; charity vaunteth not itself, is not puffed up, ⁵Doth not behave itself unseemly, seeketh not her own, is not easily provoked, thinketh no evil; ⁶Rejoiceth not in iniquity, but rejoiceth in the truth; ⁷* Beareth all things, believeth all things, hopeth all things, endureth all things. ⁸Charity never faileth: but whether there be prophecies, they shall fail; whether there be tongues, they shall cease; whether there be knowledge, it shall vanish away. ⁹For we know in part, and we prophesy in part. ¹⁰But when that which is perfect is come, then that which is in part shall be done away. ¹¹When I was a child, I spake as a child, I understood as a child, I thought as a child: but when I became a man, I put away childish things. ¹²For now we see through a glass, darkly; but then face to face: now I know in part; but then shall I know even as also I am known. [13] And now abideth faith, hope, charity, these three; but the greatest of these is charity.

Pastor Moses closed the eulogy of Rose with a prayer of benediction, and reminded the mourners, and called on all of the residents of the Mule Ridge community, that it would take the love of everyone to assist in the nurturing and raising of Rose's children.

Rose was buried in the negro section of the Mule Ridge Cemetery in a grave next to her grandparents, Andrew and Mabel Broussard, on the side of her grandfather. Her tombstone marker exclaimed: I DID THE BEST THAT I COULD!

Of note was a card of condolences signed by Benjamin.

During the following days, weeks, and months of mourning, Mary, who turned seventeen years old on August 10, 1948, and had completed her junior year of high school, was now the matriarch of the Gilmore family. She was a mother figure to both John David and Rachel.

Breaking the gloom and despair of their mother's untimely death, Mary did what her mother, Rose, had always done for her children on their birthday. Mary planned a picnic and birthday party for John David.

July 26, 1948, John David's birthday fell on a Monday, a day of work. Therefore, Mary scheduled the celebration to be held on Saturday, July 24th, during the afternoon, when everyone would be off from work.

While the Gilmore family had access to several gasoline-powered vehicles, they preferred to be transported to the picnic area of the Blue Hole riding on the family wagon pulled by their pet mule, Blue. It was a practice initiated and enjoyed by their mother, Rose.

The picnic was supplied with an assortment of food and bottled soft drinks, which included an assortment of lunch meats—bologna, headcheese, pickle loaf—canned pork and beans, hot dogs and buns with garnishments, and a variety of pastries and cookies and a chocolate cake for dessert.

Added to the traditional activities of horse shoe, sack races, and swimming, their father, Alonzo, brought fishing gear and equipment, along with cooking oil to deep fry the fish he planned on catching. However, just in case that he did not catch any fish, their father had a plan B, which was to purchase cleaned and ready-to-fry fish from the Wallace's fish business.

And, of course, Mary brought apples for Blue.

John David's birthday celebration was filled with boisterous and joyful excitement, which gave the family an opportunity to escape the hurt and pain of their mother's death. However, with the memories of Rose still fresh in the minds of the Gilmore family, Alonzo shed a tear over not being able to share the celebration with his beloved wife.

Enjoying a bright, sunny, and hot July day that was filled with the consumption of a variety of food and soft drinks, including an occasional deep fried fish sandwich, which had been caught and prepared by their father, the Gilmore family spirits were also uplifted with kind words of support and encouragement from other negro families who were enjoying the beauty, relaxation, and comfort offered from the various activities surrounding the Blue Hole, with swimming being the most popular activity.

While lounging on the bank of the Blue Hole, with a fishing pole in his hand, Alonzo heard an alarming outburst of screams.

Running toward the direction of the screams and commotion, Alonzo saw his son, John David, in distress in the water of the swimming area of the Blue Hole. The water where his son was having difficulty, was approximately 3 feet deep. John David, who would be turning seven years old within a couple of days, stood at least 4 feet 8 inches in height. But he did not know how to swim!

Wading into the shallow water to rescue his son, Alonzo, with a confident and comforting tone in his voice commanded, "John David, son, just stand on your feet."

John David, now nervously holding onto his father's legs, stood to his feet. Once standing, he looked into his father's eyes. "Dad, you need to teach me how to swim, just like you taught my sisters, and brothers."

After chastising his older children for not sharing and teaching their younger brother how to swim, Alonzo guided his youngest son into deeper water for the purpose of teaching John David on how to swim.

As he had taught each of his older children, Alonzo used the technique known as dog paddling to teach his youngest son how to swim.

"When dogs are swimming, they never use all four of their legs and feet at the same time to get them to safety. Dogs will use either their front legs and paws for movement through the water, while allowing their back legs and paws to rest, and vice-versa. Dogs always swim with the flow of the currents, while slowly angling toward the bank, and dry land," Alonzo emphasized, while noting that dogs do not seek for speed when swimming, but they swim to survive.

After teaching John David, the technique of swimming, and with the setting sun, Mary called out for Blue to come. And, after hitching the animal to the family's wagon, with the picnic's left-over food and equipment secured in the wagon, Mary gave Blue an apple as the animal transported the family to their home.

At home, and just before bed time, led by their father, the siblings thanked and blessed their older sister for the wonderful birthday party and picnic she had planned for John David, while noting that her, Mary's, eighteenth birthday would be in a few weeks, on August 10th.

When Mary requested for her birthday to be celebrated at their home, with a chocolate cake, purchased from the bakery in the city of Wyatt, and vanilla ice cream purchased from one of the local grocery stores, it signaled that the Gilmore family's love for each other was growing toward the proverbial inevitable love toward separation, which happens to sisters and brothers, as each member establishes their individual identity and personality while in pursuit of their personal dream.

The Thanksgiving and Christmas holiday season of 1948 was the first for the Gilmore Children without their mother, Rose, which added new a sting in their hearts as they celebrated a time of gratefulness and family.

Thanksgiving dinner consisted of the usual, which included a turkey and ham for meat, with an assortment of vegetables, plus a peach cobbler. The Thanksgiving dinner was prepared by Alonzo and Mary.

For Christmas, the family's meal was similar to what had been served for the Thanksgiving dinner, With Alonzo making sure that each of his children received the gift that they had requested for Santa Claus to put into their Christmas basket. Rachel, now eight months old, received a new baby buggy, as the old baby buggy, which had been used by her older sisters and brothers, had been gifted by Rose to an expectant mother of the Smithfield Plantation. John David received his first weapon. A Red Ryder, single shot, BB gun.

On the day after Christmas, John David received his first lesson on the responsibility of owning a weapon. It was a lesson that Alonzo had taught to his two older sons, Jonathan and Jacob, when they received their first weapons.

"John David, never point a gun, loaded or unloaded, at someone."

"Yes sir, Dad, because it just might be loaded."

"John David, never kill an animal just for fun and sport. Kill an animal for food, or when a wild animal presents a danger to a human, livestock, or other property."

"Yes sir, Dad. I will only kill animals for food, and to protect human life, and property," John David repeated, assuring his father that he understood the lesson.

"John David, make sure that you make the first shot a kill shot. Aim for the head. You may never get the chance for a second shot," Alonzo said, as he taught his son how to hold a long gun. "Keep both eyes open, line the sight indicator, which is at the end, and on top of the barrel of the gun, with the target, exhale, and hold your breath, while gently squeezing the trigger." After these instructions, his father, Alonzo, handed the BB-gun, with one BB, to John David.

"Yes sir, Dad!" John David responded as he loaded the single BB into his Red Ryder BB gun, and following his father's instructions, took sight on the hand-made target, and recorded a bull's eye hit from approximately seventy-five feet away from the target.

"Great shot, son!" shouted a proud father, while congratulating his son, and pondering in his heart that his son was a natural marksman.

With the Gilmore children, under the leadership of their father, maturing, growing in age, and knowledge, as well as personal independence, life began to be measured with graduation from elementary school, a driver's license, graduating from high school, and going away to college.

In May of 1951, at the end of the school year for the negro segregated high school of Wyatt, Missouri, Jonathan became the first of the Gilmore children to complete the twelfth grade. He graduated with honors from a class of nine.

For the following year, Jonathan was a tremendous help to his father in farming the Smithfield Plantation as it was transforming from mule power to gasoline-powered tractors and other mechanized equipment.

In February of 1952, the month of his nineteenth birthday, Jonathan was drafted into the United States Army, as an infantryman.

On a day in late June of 1953, just before the end of the Korean War, and during the Battle of Kumsong, Corporal Jonathan Gilmore was killed in action. He was twenty. Jonathan was buried in the United Nations Memorial Cemetery, located at Tangguh, in the Nam District, City of Busan, Republic of Korea.

Earlier, on a happier occasion, at the end of the school year of 1953, Martha graduated as valedictorian of her class of thirteen. She was accepted, on a full academic scholarship, to attend Tuskegee University of Tuskegee, Alabama. Martha's dream to become a school teacher was possible.

Jacob completed high school in May of 1955. He served as president of his senior class of fifteen, and led his class to the Negro Mathematics Championship for the State of Missouri. While receiving numerous scholastic scholarship offers to attend predominantly negro colleges and universities, Jacob successfully applied for and received a military deferment so he could assist his father on the Smithfield Plantation.

Sarah, as it was purported in the adult male community, was exclaimed to be the proverbial "spit and image" of her mother, Rose. Sarah was described as a "sexually drop-dead beauty." Upon graduating from high school in the spring of 1957, and with permission and blessings from her father, Sarah was allowed to move to Hollywood, California to pursue her dream to become an actress.

Sarah, because of her uncanny resemblance to her mother, Rose, whom her Nanna loved dearly, was her great-grandmother's favorite. She received unlimited financial support from Mrs. Smithfield, who managed and supervised Sarah's move to Hollywood.

In the early afternoon, on the first Saturday of June 1957, Sarah was driven by her father, along with her brothers, Jacob and John David, and sisters Mary and Rachel, to the St. Louis Lambert Airport of St. Louis, Missouri. After tearful goodbyes between the family, Sarah boarded Trans World Airline (TWA) flight number 726 to Hollywood, California. The plane was a Lockheed Constellation, propeller-driven, four-engine aircraft. Sara's dream to become an actress and movie star was now becoming a reality.

The car ride home for the remaining members of the Gilmore family proceeded in quiet and somber reflections, as the family was growing smaller and toward separation.

In early 1958, while Sarah was in the process of building her resume and establishing her career as an actress, Mrs. Smithfield contracted for the Smithfield Plantation to become a licensed corporation of commercial farmers, and returned to live in her birth home of New Orleans, Louisiana.

Within the contract were two conditions that Mrs. Smithfield demanded: (1). As long as a male heir, with the surname of Gilmore, could farm, he must have an option to purchase five hundred acres of his choice of the Smithfield Plantation at the 1958 cost per acre. (2). A life insurance policy of $10,000 for each child of Alonzo and Rose Gilmore be purchased with profits from farming the Gilmore Plantation, with the beneficiary being the oldest family member alive.

In the meantime, Mary, with the burden and responsibility of being the matriarch and mother figure to Rachel and John David, who had begun to address their older sister as "Mama Mary," found comfort and companionship by consuming alcohol.,

While working as a maid and housekeeper in her great-grandmother's home, Mary discovered Mrs. Adaline's well-stocked wine cellar, and introduced herself to the sweet elixir, which gave to her a moment of peace, relaxation, and comfort.

John David graduated from high school on the last Friday of May, 1959. He was both class valedictorian and president of his class, which numbered fourteen.

Chapter Fifteen

Never To Return Home

(THE MIDDLE YEARS)

I t was early in the afternoon, on the first Saturday of September, 1959. John David Gilmore was eighteen years old, and had never spent a single day away from his family, especially his father, Alonzo, whom John David looked up to as his hero.

John David and Rachel, being the two youngest siblings, had become very close in their relationship, and protective of each other. Rachel had begun to address John David as "Big Brother". To Rachel, the baby of the Gilmore family, John David had become her mentor and guardian.

Before departing for college, with the promise of secrecy, John David and Rachel discussed their concern for Mary's growing consumption of alcohol, and they wished for their dad to send her to Hollywood, California to live with their sister, Sarah. During their conversation, John David promised Rachel that he would make a telephone call to Sarah, and inform her of Mary's dependence on alcohol, and how her drinking was affecting her health.

In the meantime, their father, Alonzo, was acutely aware of his oldest child, Mary's, growing addiction for alcohol. Alonzo and Sarah had decided that it would be best for both Mary and Rachel to move to California to live with Sarah.

It was bitter-sweet for John David to leave his family and home to enroll at the University of Missouri, on a full-paid, room and board, academic scholarship. To lessen the sadness of leaving home and family, John David persuaded his father to allow him, alone, to take the Greyhound Bus to Columbia, Missouri, the campus of the University of Missouri.

With the two-piece, brown-leather luggage set that Mrs. Adaline had gifted him with upon his graduation from high school, and containing all of his clothing and personal items, John David embraced his father, Alonzo, brother, Jacob, and baby sister, Rachel. With the promise that he would make a telephone call upon arriving at his campus dormitory to let his family know that he was safe. John David, at the Wyatt, Missouri bus station, said goodbye to his family and boarded a Greyhound bus. Sister, Mary, who was inebriated from drinking wine, chose to remain at home.

"As I made my first step on to the Greyhound bus, I knew within my mind, heart, soul, and spirit that I would never return to my birth place, home and my sisters, brother, and father. I would never see and embrace them again. From that day forward, to survive the challenges of life, I would have to rely on all that my father taught me, which included how to shoot a rifle, how to swim, how to butcher wild game, and how to court a woman and satisfy her in love making", words that John David shared with me with prideful humility.

After a six-hour bus ride, John David arrived at the Greyhound bus terminal in Columbia, Missouri, where he, and hundreds of other students were transported to their respective dormitories by campus shuttle buses.

After being checked into his dormitory room by university personnel, John David excitedly made a telephone call to his father, insuring him that his son was safe.

It was in the early morning hours of September 5, 1959 when John David left his assigned dormitory room, equipped with a campus guide, and map of buildings and their locations, to have breakfast at the designated campus cafeteria that served his dormitory. After breakfast, which consisted of cold cereal, a variety of fruit, milk, and toast, John David proceeded to the assigned building for freshman enrollment and registration.

After enrolling for the required courses to receive a bachelor of science degree in mathematics, which included a choice of one of the branches of the United States Military, John David chose the Air Force Reserve Officer Training Corp (AFROTC).

John David's first semester class registration and enrollment included:

- Algebra 1 (5 semester hours).

- Physics 1 (5 semester hours).

- Geometry 1 (5 semester hours).

- Freshman English/Essay Writing (3 semester hours).

- Freshman AFROTC (2 semester hours).

Upon completing his class registration and enrollment, John David was directed to proceed to the campus book store, where, never having new text books, he went to the used book section to select his required text books and study materials.

The cashier at the checkout counter displayed a friendly smile. "John David, you are attending the university on a full-paid academic scholarship, which entitles you to select new books," he advised.

Because John David's early years of education had been administered through segregated school systems in southeast Missouri, and had been supplied with used books from the white segregated schools, John David had assumed that he had to select used books to complete his class assignments. This was the first time for John David to own new books.

Due to his financial hardship, John David received permission from the university's administration to remain on campus during the Thanksgiving and Christmas holiday school break. During the holiday break, the campus cafeteria was closed; therefore, the students who chose to remain on campus were responsible for their daily food requirements.

To provide food for himself, John David recalled the experiences that his father, Alonzo, and Alonzo's brothers Benjamin and James had on their travels from the Gilmore Plantation. To survive, the brothers would hire themselves out as laborers in exchange for food and a place to sleep. John David used this lesson to go into the communities that surrounded the campus and introduced himself to the home owners, and that he was a student at the university who did not have the necessary finances to travel to his home, and that he was prepared to work for food.

His offer was immediately accepted. He worked to wash and detail automobiles, rake leaves, lawn care, and clean and organize automobile garages. For his labor, John David was supplied with an abundance of food, financial tips, and invited to be a guest for the various families' Thanksgiving and Christmas dinners.

The financial payments that were made to John David, he sent to his dad to help in the purchase of the farm, which was a dream of his father.

The 1959 Thanksgiving and Christmas holiday celebrations for the Gilmore family were conducted via telephone conversations. This style of celebration and communication would become the norm for the Gilmore family for the future years. For the 1959 holiday season, and on the evening of Thanksgiving, the Gilmore family exchanged greetings, and news updates by using a telephone conference call.

"Greetings to all of my children," Alonzo said. "Mary and Rachel have agreed to move to Hollywood, California to live with Sarah. The relocation will be carried out in early June of 1960, when school, for Rachel, is ended for the cotton chopping season. We are very close, with our savings plan, to having enough money to purchase the five hundred acres of farm land that your mother's Nanna, Mrs. Smithfield, promised to sell to the Gilmore family. Finally, your dad's health is beginning to fail!" exclaimed the father, with a tone of lonely sadness in his voice.

Mary replied, "Happy Thanksgiving, and Merry Christmas to my dear sisters, and brothers. Rachel and I are so looking forward to moving to California. Sarah, thanks so very much for allowing us to live with you. I am drinking too much wine, and a new environment will help me to kick this alcohol addiction!" Mary confided, with a shaky voice.

"I am so missing home," Martha said, "and being with my family. I am wishing each of you a Happy Thanksgiving and a Merry Christmas. I have received my teacher's license to teach high school mathematics and science. And Dad, I will be sending a check to help with the purchase of the farm. Remember the vow that we made to Mom, that we would always support each other, and follow our dreams," Martha concluded with a solemn tone in her voice.

Rachel jumped in. "Hello to my big brothers, and big sisters. I will be twelve years old on April 22, 1960. I am making As in all of my classes, and I will be graduating from the eighth-grade at the end of May. I have decided to become an Obstetrician Gynecologist (OBGYN) when I go to college. I want to put an end to negro women dying while giving birth to their babies, as our mother did, giving birth to me," Rachel vowed.

"Happy! Happy! Thanksgiving and Merry Christmas to all of my family!" Sarah said. "For Mary and Rachel, your rooms are ready! I am so looking forward to having each of you to come and live with me. Great-grandmother Adaline has helped financially to provide us with a wonderful home that has three bedrooms and two bathrooms, with showers. For Rachel, our mother's Nanna has made arrangements for you to be enrolled in a prestigious private school. I am so missing home, and my family!" Sarah sniffed. "And more acting opportunities are being made available to me, so before I begin crying, goodbye," she said and reluctantly hung up the telephone.

John David spoke next. "I am wishing each of my sisters, and brothers, and especially my dad a Happy Thanksgiving and a Merry Christmas. I am doing well with my college studies. I have a straight-A average, and I will be on the dean's list for academic excellence. I have been invited to attend advanced AFROTC training during the summer break; therefore, I will not be able to return home this summer. During the holiday semester break, I plan on working as a laborer to provide myself with food. And whatever cash earnings I may receive, I will contribute to the savings account so that Dad will be able to purchase the farm," John David concluded with a tone of maturity and loneliness in his voice.

"Hello family!" Jacob said. "I am well, and our pet mule, Blue, is well. Both of us are wishing each of you a blessed holiday season. Do not forget the promise that we made to our mother 'that we would always support each other, listen to our heart, and follow after our dream,'" Jacob reminded his sisters and brother. "Love for family always grows toward separation, but we are still family," he concluded with a tone of melancholy in his voice.

Chapter Sixteen

The Making of An Asset

I n June of 1961, John David Gilmore completed his sopho-more year of college as he pursued a Bachelor of Science De-gree in Mathematics. With an A grade-point-average, and academic recognition from the dean of mathematics, and the commander of the AFROTC, John David was recommended to be interviewed by Brigadier General Thornton "Thorny" Drinkwater, recruiter for the United States Air Force's Strategic Air Command (SAC), Military Intelligence Division.

In the meantime, on June 15, 1961, John David received a Western Union Telegram informing him that his sister, Mary, had died, and for him to make a telephone call to his father as soon as possible.

Alonzo explained to his son that his oldest sister, Mary, had lost her battle with alcohol, and had died from alcohol poisoning and sclerosis of the liver.

"Due to my failing health, and dependance on your brother, Jacob, to help in taking care of me, as well as taking care of the farm, Jacob and I will not be able to attend your sister's funeral," Alonzo, a tearful father, explained to his son.

Mary Gilmore was born on August 10, 1931, and died on June 15, 1961. She was 29.

Mary left to mourn: her father, Alonzo; two brothers, Jacob and John David; and three sisters, Martha, Sarah, and Rachel. She was preceded in death by her mother, Rose, and brother, Jonathan.

At a grave-side service, attended by sisters Sarah and Rachel, in the Hollywood Cemetery of California, Mary Gilmore was buried in the afternoon on June 22, 1961. At her request, Mary's tombstone read: "MOM I AM SORRY. I FAILED!"

Flowers, and cards of condolences from Mary's father, sisters, and brothers, conveyed their grief, with explanations of health, or employment responsibilities. They were sorry that they could not attend the home going celebration.

On a bitter-sweet note, the proceeds from Mary's ten-thousand-dollar life insurance policy were enough cash to allow her father, Alonzo, to purchase the five-hundred-acre farm at the 1958 cost per acre, for a total cost of $51,780 dollars.

So, in death, Mary, with the proceeds from the life insurance that was purchased by her great grandmother, Mrs. Adaline Smithfield, was able to keep her vow that she made to her mother. "To always support the family."

It was a Saturday, July 1, 1961, when John David was interviewed by Brigadier General Thornton "Thorny" Drinkwater, who arrived from his Washington, DC assignment at the Pentagon to conduct the interview and evaluation of John David. The interview was conducted at the office of the Joint Reserve Officer Training Corp Command Center, which was located on the university's campus.

At 10:00 a.m., after politely knocking on the office door, John David was instructed to enter, where he saluted and introduced himself to Brigadier General Drinkwater, who stood 6 feet 3 inches in height, weighing approximately 190 pounds, with deep-set brown eyes, and bushy eye-brows. The general's dark brown hair was cut very short, which accentuated his square-jawed face, chiseled chin, and pleasant-sized pointed nose. He was a perfect model for a poster to advertise to recruit enlistment into the United States Military.

The interview began when General Drinkwater turned his back to John David and asked him, "What is the sum of two plus two equal?"

John David, remembering a lesson that his father had taught him: Always carefully listen to the question, but, also, pay close attention to who is asking the question.

"General, sir, two plus two equals whatever you want it to equal!" John David exclaimed with a tone of confidence in his voice.

Turning to face the interviewee, General Drinkwater was pleasantly surprised with the confidence that John David displayed in his response, and answer to his "gotcha" question. The general motioned for John David to have a seat, indicating that the interview would continue.

"John David, I have interviewed dozens of applicants, and you are the first to respond with the correct answer, which is functional. Are you as smart as your resume, and academic transcript suggest?" inquired the General.

John David recalled another lesson that his father had taught him: Keep the answers to all questions short

"I am sure that you will determine how smart I am," John David responded with a tone of quiet confidence in his voice.

"John David, how did you know that I was searching for a functional answer to the mathematical equation of what is the sum two plus two?" General Drinkwater inquired, with a quizzical tone in his voice.

"Sir, you would not have achieved the rank of Brigadier General if you did not know that two plus two equals four. I believe that you are looking for a man with the character and skills to get the job done. Regardless. My sister, Martha, taught me arithmetic addition when I was at the age of three, therefore, logically, there was just one answer—a functional answer," John David calmly responded.

General Drinkwater said, "To your response—absolutely. Also, I see here on your marksmanship score card, that you always choose a single shot weapon, over weapons with a magazine, and having multi-shot capability. Why?" inquired the General, as he rose from his chair, and positioned himself to stand over John David with the intent to see if he could intimidate the young cadet.

"When I was six years old, sir, my dad, Alonzo Gilmore, taught me gun safety, and how to shoot a long gun, using a single shot Red Ryder BB gun. Because we hunted, and depended on wild animals for food, it was a necessity to make the first shot a kill shot, because no wild animal would give you a second shot to end its life. These lessons included the sport of clay pigeon, skeet shooting, with a single shot, 20-gauge-shot gun. To accomplish expert marksmanship at clay pigeon, skeet shooting, first shot success was a necessity," John David exclaimed, with a tone of command in his voice.

General Drinkwater slowly nodded. "Amazing. Now to your swimming skills. Your records tell me that during your survival training, which was conducted during ROTC summer camp, that you set a record of five-hours of swimming endurance. The previous record was three-hours and fifteen minutes. Explain to me on your record setting accomplishments," General Drinkwater commanded.

"Sir, again, my father, who is my hero, taught me how to swim. The teaching model that he used was fashioned after how dogs swim, which is called dog paddling. Dogs' natural instincts are for them to swim for survival, not for speed. Endurance is the key to their survival; Therefore, dogs never use all four of their paws and legs at the same time. When they use their front paws, and legs, their back legs, and paws are at rest, and vice versa. This technique conserves energy, which is necessary for endurance. In addition, dogs swim with the currents, while slowly angling toward the shore line to increase the chances for survival," John David explained, while noting that the instructor for the swimming session had put an emphasis on endurance, and not speed, while advising the participants to pay attention to the changing direction of the currents. "I remembered to use the lessons that my father taught me," concluded John David.

Placing his hands behind his back, and with his notorious 'Thorny' stare in to the eyes of John David, General Drinkwater said, "Mr. John David Gilmore, I have never been so impressed with an applicant as I have been with you. I am authorized, by the Secretary of Defense, to offer to you a training position into the United States School of Special Operations of Military Intelligence. John David, if you accept this assignment, you will continue with your studies in pursuing a Bachelor of Science Degree in Mathematics, as well as top secret training to include but not limited to becoming a fighter jet pilot and a trained sniper. The training will be for a minimum of two years, and you will not be permitted to break training for any reason, to include attending funerals of loved ones. John David Gilmore, do you accept this offer?"

"Yes, sir, General Drinkwater. I accept the offer, and I understand the conditions," replied John David confidently and with a tone of authority in his voice.

"Wonderful!" General Drinkwater said. "Mr. Gilmore, you will never regret the decision you have made today. You will become an invaluable asset to your country.

Go to your dormitory, and pack your luggage. Make a telephone call to your father, and inform him of the decision that you have made," instructed General Drinkwater.

"John David," continued the General, "a military attaché will arrive at your dormitory, tomorrow morning, Sunday, to provide transportation to where you will be trained, and I will be personally monitoring your training, and progress," concluded the General.

After departing with a salute and a goodbye to General Thornton Drinkwater, John David made a telephone call, informing his father, Alonzo, and brother, Jacob of his opportunity for top secret training.

Father, Alonzo, and brother, Jacob were the last to remain in the home of the Gilmore family.

Alonzo and Jacob encouraged and ensured John David that he would be successful. Both father and brother, with a tone of bitter-sweet emotions in their voices, said, "Goodbye," and noted that they would inform his sisters of their brother's opportunity to become an asset for his country.

Chapter Seventeen

Top Secret – Making of An Asset

I t was a Monday, July 3, 1961, approximately six o'clock, on a warm sunny morning, when John David departed from his one-bedroom, kitchenette-efficiency apartment. Enroute to the campus cafeteria for breakfast, John David became acutely aware of the early morning heavy humidity, high temperature, and high blue, cloudless sky. Also of note were the trees and shrubbery. John David surmised that he was somewhere in the state of Florida.

At 8:00 a.m., John David's schedule required for him to report to the campus medical center to receive a physical examination before reporting for official orientation to be admitted into The United States of America Special Operations Academy. Better known as the Academy.

John David's physical reported him to be:

- Age: 20 (DOB: 7/26/1941)

- Race: Negro, with high light-brown skin

- Hair: Black, and wavy

- Eyes: Grey (having excellent 20/5 vision)

- Height: 6'-2"

- Weight: 170 pounds

- Blood Pressure: 118/78

- Dental: Good

- General Health: Excellent

After completing his required physical examination, John David proceeded to the hall designated for orientation, where he was introduced to six additional trainees, totaling a class of seven.

Colonel Wade McCleary, Military Intelligence Officer of the United States Air Force, was the superintendent of training at the campus. After introducing himself, Colonel McCleary directed the trainees to open the sealed top-secret envelopes that were on their desks, which contained the rules and regulations that would govern each trainee:

- The duration of the training will take approximately twenty-four months.

- There will be no off-campus vacations. No exceptions.

- Trainees will be trained in the principles of Aeronautics, Aircraft Mechanics, Piloting of latest models of Fighter Jets, High Altitude Reconnaissance Jets, and Rotary Wing Air craft (Helicopters).

- Trainees will be trained and certified as Low and High-Altitude Paratroopers.

- Trainees will be trained in the use of Offensive and Defensive Sniper Weaponry.

- Trainees will be trained and certified as Self Contained Breathing Apparatus (SCUBA) Divers.

- Trainees will utilize only those telephones that have been designated as secure.

- Trainees will not discuss the details of their individual training with anyone, to include fellow trainees. THIS TRAINING IS TOP SECRET!

- Trainees will be assigned a college professor, who will assist and tutor each trainee to successfully satisfy the academic requirements for the college degree that each trainee is pursuing, and a foreign language of the trainee's choice.

- Each trainee will be compensated at the pay-grade of a military E5, which is $145.07 per month. These finances, or any portion, may be paid directly to a beneficiary of the trainee's choice.

- All necessities, including food and clothing, will be provided by the United States Government. A credit card is in your folder, and has been issued in each trainee's name, and may be used at the discretion of the trainee.

- There is zero tolerance to these rules and regulations. Violations will result in immediate termination of training, and dismissal from the campus.

"I wish each trainee the best, and if a trainee ever needs my help, remember that I have an open-door policy. Appointments are not necessary." Colonel McCleary concluded with his orientation by noting, "Always listen carefully to all instructions, and remember that to solve certain problems, a trainee may have to change the equation."

After the orientation, John David, utilizing a secure telephone, made a call to his father, Alonzo, and brother, Jacob, informing them what he could expect during the next two years that he would receive, while pondering in his heart what was the meaning of the comment made by Colonel McCleary: "To solve certain problems, a trainee may have to change the equation."

From July 1961 to July 1963, John David's training schedule included, but was not limited to classroom lectures on the principles of flight and aeronautics, aircraft mechanics, hands-on instructions with the aircrafts, which was supervised by fighter jet instructor pilots, seasoned and experienced weaponry and sniper instructors, SCUBA divers, and swimming instructors. These lessons, instructions, and training were in duration of ten to twelve hours per day, for five and a half days each week. Trainees were allowed to have Saturday afternoons and Sundays off from training for the purposes of recreation, worshipping, and the management of their personal affairs.

On Monday, October 1, 1962, while attending a class room lecture and receiving instructions on the principles of how to pack a parachute, and high altitude sky-diving, John David received orders to report to the office of the campus superintendent, where he was informed that his father, Alonzo Gilmore, had died.

Using the secure telephone in the superintendent's office, John David made a telephone call to his brother, Jacob, who informed John

David that their father had died from the disease of stomach cancer, and their father would be buried on Saturday, October 6, 1962.

"Your sisters and I understand that you are not allowed to break training, and Dad would understand too; therefore, *we* will take care of the funeral arrangements," Jacob exclaimed. "John David," continued Jacob, while consoling his younger brother, with a quiet assurance in his voice, "Dad would want for you to not break with your training, but to press on. Everything will be alright, and Dad would want you to be strong," Jacob said.

"I will Jacob, um, I will be strong and press on. Thanks. It's nice to hear your voice. Goodbye."

On Friday, October 5, 1962, in the afternoon of a bright, warm, and sunny fall day, John David's sister, Martha, arrived at the Gilmore family home after an eight-hour automobile drive from Tuskegee, Alabama, her current home and city of employment as a high school teacher.

While Jacob greeted Martha, and assisted her with her luggage, Sarah and Rachel arrived with Sarah driving a rental automobile, which they had secured after arriving at the Saint Louis, Missouri, Lamber Airport, on a TWA flight from Hollywood.

After greetings and heartfelt embraces, Jacob reminded his sisters that they knew where their bedrooms were, and asked if they had brought with them, as he requested, their tape recorders, because he planned for them to record their conversations for the purpose of sending to John David the dynamic and interactive real-time recordings of the events surrounding their father's funeral.

"Yes, I brought my tape recorder, and a supply of batteries," replied Martha.

"Me too."

"And so did I," echoed Sarah and Rachel.

"Well, then, turn your recorders on, now," Jacob commanded with laughter.

"Bossy, bossy. Just like Dad," Sarah responded with a wry chuckle in her voice.

"And don't forget how controlling and demanding Mom was," Martha added.

"Whatever," Jacob said. "Okay," he continued, "the recording of our conversations, and reminiscing will begin at this very moment."

"Amen," added Rachel, with a loud outburst of laughter.

Recorded Interactive and Dynamic Conversations of Siblings: Jacob, Martha, Sarah, and Rachel.

(Friday, October 5, 1962)

"Mom's sewing room and sewing machine is in the same position that "Mama Mary," and I left it in when we moved to Hollywood to live with Sarah. That's the gown she was making for me before I was born," mused Rachel.

"Yes, it is pink!" Sarah added, with a choking tone in her voice. "Mom knew that her next baby would be a girl, because I was the last girl, before John David was born, and we were all born in the order of 'girl, boy, then girl.'"

"And Dad made me promise that I would always leave it the same, as it was when she died," Jacob said, while clarifying why things were the same.

"I, so miss, Mama Mary! Why did she have to drink so much alcohol?" inquired a teary-eyed Rachel.

"I don't know. You have been asking me that same question every day, since she passed away! Stop asking me that same question!" Sarah replied to her baby sister, in a testy and impatient tone of voice.

"We have an abundance of food. The neighbors, beginning with when Dad became home-bound, have been so kind and generous.

They are always bringing food. I am now preparing and setting the dining room table. We can eat now, and continue to reminisce while we are eating," Jacob suggested, as he completed setting the table.

Sarah, noting that she was just eighteen months younger than Jonathan, and were rival siblings, said, "I was just wondering if the Veteran's Administration will help us to bring Jonathan's remains from Korea, to be buried alongside Mom and Dad."

"I think so," Jacob replied. "Dad and I were working on that before he passed away. Dad would so love for that to happen!"

"Speaking of Dad," Martha interjected, "I noticed that there is a sleeping cot in the living room. Did he stop sleeping in his and Mom's bedroom?"

"Yes," Jacob recalled. "Dad stopped sleeping in that bedroom the very same day that we buried Mom. Dad died on that cot. We have a busy day before us tomorrow. I am going to take a shower, and go to bed to get some sleep. Good night to you girls. See you tomorrow morning at breakfast. We are scheduled to be at the Mule Ridge Cemetery at ten o'clock tomorrow morning."

Saturday—Conversations Continue

During breakfast on Saturday morning, which included bacon and eggs, grits, toast, milk, orange juice, and coffee, Rachel reminded her older sisters, and brother, to turn on their tape recorders.

"And because I am the youngest child, do y'all think Blue will remember me?" Rachel asked.

"Yes, he will remember you," Sarah said.

"Take an apple, and feed it to him. He is probably standing outside of Mom's bedroom window. He does that every morning," Jacob said.

"How old is Blue?" Rachel asked.

"Blue is thirty-one," Jacob replied.

"Wow! The same age as Mama Mary would be, had she not died from alcoholism. Correct?" Rachel said.

"Mules can live to be forty-five to fifty years, especially if they are fed well and given good veterinarian care, which Blue has received. Blue has always been a pet, and not a work animal," Martha said.

"I will be doing the driving, and we will take the rental car. Rachel, sit with me in the front passenger seat, and allow your sisters to sit in the back seat, and we will proceed to the cemetery," Jacob instructed.

"There's the pecan tree where I sat with John David and taught him how to read and do simple arithmetic," Martha said. "I always dreamed to become a school teacher, and I would practice my skills by teaching John David. He would always complain that I was a hard and no-nonsense teacher. I will soon be receiving my master's degree in education, and have been offered a position of associate professor at Tuskegee University, when I graduate."

"Wonderful!" Sarah exclaimed. "Rachel will be graduating from high school before her sixteenth birthday. She is a straight-A student, and will be allowed to enroll at the University of California at Los Angeles (UCLA) in the school of premed."

"And I promise each of you that I will become an OBGYN medical doctor! Did y'all know that Sarah has been offered to be the leading actress in a soon-to-be *big time* movie?"

"Contracts have yet to be signed," Sarah interrupted, "but my agent tells me that I am at the top of the list for the part!"

"We have arrived at the Mule Ridge Cemetery," Jacob announced, "and we thank God for a beautiful and sunny day."

"We can all say Amen to that," Martha said as the three sisters joined in with Jacob, thanking God for a beautiful day.

The Gilmore siblings, with Jacob attired in a black two-piece suit, white shirt, and black neck tie, along with his sisters, who were attired in black dresses and black slippers, greeted and thanked the funeral directors, and proceeded with the grave-side service as requested by their late father.

The Obituary:

Alonzo Gilmore

February 29,1904? - October 1, 1962

- Prayer (in unison, all the siblings)—Prayer of Jabez: 1 Chronicles, Chapter 4: 9-10.

- Scripture (in unison, all the siblings)—23rd Psalm.

- Hymn (sung by all the siblings)—Amazing Grace.

- Words of comfort (in unison, all siblings)—Vow Made to Your Mother: "We Will Always Support Each Other!"

- Alonzo Gilmore was born on the Gilmore Plantation, in Tupelo, Mississippi, to Leah Gilmore, a single mom. She preceded him in death. He was married to Rose Broussard Gilmore. They were married for nineteen years. She preceded him in death. They were the parents of four daughters and three sons. The oldest son, Jonathan, and oldest daughter, Mary preceded him in death. Alonzo Gilmore is survived by, in order of birth, Martha, Jacob, Sarah, John David, and Rachel Gilmore.

- Benediction (in unison, all the siblings)—The Lord's Prayer: Matthew Chapter 6: 9-13.

Alonzo Gilmore was buried alongside of his wife, Rose. His tombstone read: TOGETHER AGAIN—FOREVER!

Conversations Continue on The Ride Home

"Since y'all have been away from home, Pastor and Mother Moses have passed away, and that is the reason why Dad requested for us to have a grave-side funeral for him," Jacob began back with the conversation on their return home. "The Mule Ridge Baptist Church has been decommissioned, and torn down. Also, the hut that Dad and Uncle Benjamin lived in, plus the home of Pastor and Mother Moses lived in has been torn down, and the land is being used to grow soy beans and corn."

"May we go to the Blue Hole?" Rachel said. "I want to see where Dad taught John David how to swim, after he almost drowned."

"Yes, Rachel, I will drive down to the Blue Hole."

"Dad also taught you, Rachel. Matter of fact, Dad taught all of us how to swim at that same location," noted Martha.

"That's right," Sarah added, while recalling the many, and wonderful good times they had at the Blue Hole, including their pet mule Blue.

Happy laughter from all the siblings filled the automobile. Jacob reminded his sisters that all of them were baptized into the Christian Faith by Pastor Moses at the Blue Hole.

Let us all say, "Bless his soul," Rachel added.

Back Home and Conversations Continue

Arriving back to their home, the bereaved siblings were uplifted when they saw Blue standing at their mother's bedroom window when Rachel ran inside of their home to get an apple to feed to the loyal sentinel.

"Blue, you do remember me," Rachel shouted.

While sitting at the family's dining room table, and enjoying left over foods of cakes and pies that had been so generously provided for the Gilmore children, Jacob continued to bring his sisters up to date of the changes to the Countrypolitan of Wyatt, Missouri, once the economic engine for the entire region of some one-hundred square miles of many cities, and villages. Cotton production, which was the king of the region's economy, was down by approximately seventy-five percent. Farmers were now growing crops of corn, soy beans, and wheat, all requiring less human labor. "There used to be three cotton gins, now there is one. Passenger trains and commercial bus services have ceased operation to and from Wyatt. During the economic boom years, there were five grocery stores that were providing food for the population in the region. Now there is two. Dr. Fairweather relocated his seven-bed clinic to Charleston, Missouri. He has since passed away, and the clinic is now a nursing home, which is managed by his daughter, Florence, R.N. The Southeast Missouri School District is no longer segregated into white and negro. It has been integrated into Charleston School District."

"Should I leave Tuskegee, and move back home to teach?" Martha asked, in a jovial tone.

"Mama's Nanna, Mrs. Adaline, our great grandmother, mailed a condolence card to our home, and Uncle Benjamin also sent a card of condolence. Wow! Mrs. Adaline has enclosed a cashier check in the amount of $35,000!" Jacob exclaimed.

"WOW!" the siblings shouted in unison.

Jacob noted that Uncle Benjamin's card of condolence contained a personal check in the amount of $1,000. "This is enough cash for us to purchase a grain-combine to harvest the corn and soy bean crops. We will no longer have a need to hire contractors to harvest our crops. Now I can appreciate how happy Dad was when we received the proceeds from Mary's life insurance policy, which allowed Dad to purchase the farm!"

In unison, the siblings thanked and praised God for the financial blessing. And all the siblings agreed to go to their rooms, to rest, reflect, and prepare to return to their homes and every day routines after a night of sleep.

Sunday Morning Conversations and Goodbyes

At six o'clock Sunday morning, the Gilmore siblings were awakened with a telephone call from their brother, John David. Gathered around the telephone, which was placed on speaker mode, the voice of John David shouted, "Good morning, Martha! Good morning, Jacob! Good morning, Sarah! And good morning, baby sister Rachel, who will be graduating from high school at the age of fifteen, and soon thereafter will be entering into the school of premed at UCLA. I am so proud!"

"I'm the youngest, therefore I will be the first to speak! Yes, I will be enrolling in premed at UCLA, and as I promised you when you went away to the University of Missouri, I will become an OBGYN medical doctor, and find a medical solution to prevent negro women from dying while giving birth to their babies! One more thing, John David, are you training to become a spy, and work for the Central Intelligence Agency (CIA)?"

"No comment! And if I answered that, baby sister, I would have to kill you!"

Laughter from the sibling filled the ongoing conversation.

"I will soon be receiving my master's degree in education, and have been offered a position of Associate Professor at Tuskegee University!"

"I am so proud!" John David said to Martha. "And thanks so very much for teaching me how read, write, and do arithmetic. You were the best teacher I have ever had.

Sarah spoke next. "I am now receiving a fair number of roles, and I am the leading candidate for a lead role in an upcoming major movie. When I sign the contract, I am going to help Jacob to purchase some additional farming equipment. Before Dad passed away, he told me that you were contributing most of your monthly paycheck to help with the farm. So thoughtful, and caring! Mom and Dad would be so proud of us, as we continue to obey their teaching, that we should always support each other."

"After Mom's death, Dad was never his old self," Jacob said. "They loved each other *so* very much. Dad prepared his obituary, order of service for his funeral, and the final words on his tombstone, which are: TOGETHER AGAIN—FOREVER! Mom's Nanna, and our great grandmother, blessed us with a $35,000 cashier check, and Uncle Benjamin gave to us a personal check for $1,000, and on tomorrow morning, I will be purchasing a harvesting combine machine! Now, Blue and I can compete with the other farmers!"

Smiles and laughter from the other siblings.

"Blue is well, and is standing outside at Mom's bedroom window. I will soon be sending to you a copy of the conversations that we have been recording." Jacob said.

"Goodbye! Goodbye! Goodbye!" the girls all said.

"And goodbye to each of my sisters, and brother!" Jacob concluded.

So, on that chilly October Sunday morning, as the sun was rising to meet a sunny, blue, fall sky, with their luggage secured into the trunks of their automobiles, the Gilmore siblings, while being reminded by Jacob of what their father had promised, that this was home, and there would always be a bed to sleep in and food at the table to eat if any family member, for whatever reason, wished to return to their home.

With tearful embraces, and memories of their brother, Jonathan, and sister, Mary, who preceded their siblings in death, the Gilmore children said goodbye.

Now standing on the front porch of the Gilmore home, with Blue standing not far away, Jacob watched his sisters drive away, down the gravel road that they had walked when it was a dirt road, and pass the giant pecan tree with its colorful red, yellow, and orange leaves of the fall season, and where the Gilmore family enjoyed so many picnics, as they gathered the falling pecan nuts.

Jacob, now alone in the family home, pondered in his heart, *Will there ever again be a family reunion, and picnic, if not for a funeral?*

In the meantime, John David was in the final months of training at the School of Special Operations, and with a steady flow of announcements of his siblings' accomplishments, the days, weeks, and months seemed to fly by:

- On October 15,1962, Sarah signed a lucrative contract to star as the leading lady in a major movie, which allowed her to provide finances so that Jacob could purchase needed farm equipment.

- A week later, at the peak of the corn and soy-bean harvesting season, Jacob purchased a Ford Tractor-Trailer Truck and employed a licensed commercial driver to transport the harvested crops to market.

- During the week of Thanksgiving, Martha accepted a position of Associate Professor at Tuskegee University, and entered into its program to pursue a Doctorate of Philosophy (PhD) in Education.

- For his Christmas gift, John David received a copy of the recorded conversations, and a copy of his father's obituary. John David would listen to the recorded conversations daily, and confided to Jacob, "The conversations inspired me to continue and finish with my Special Operations training!"

• In March of 1963, Jacob signed a contract for builders to construct silos to store soy beans and corn, which allowed him to sell the grains when the profits were at their highest.

• On April 25, 1963, Rachel received official notice that she had been accepted at UCLA into its School of Premed. Rachel was fifteen.

• On June 25, 1963, John David was informed by Colonel Wade McCleary, Campus Superintendent of Special Operations Training, that he had successfully satisfied all training in fighter jet, high altitude reconnaissance aircraft, rotary wing (helicopter) aircraft, low- and high-altitude paratrooper certification, sniper weaponry, self-contained breathing apparatus (SCUBA) certification, foreign language (Spanish), and a bachelor degree (BSc) in mathematics.

"Also, during your two years of intense physical and academic training," Colonel McCleary said, "you are graduating number one of your class of seven. John David Gilmore, you will receive a commission of second lieutenant into the United States Air Force on July 1, 1963 at 1300 hours at the campus assembly hall.

So, on July 1, 1963, at 1300 hours, recently promoted Major General Thorny Drinkwater, of the Strategic Air Command (SAC), Military Intelligence Division, asked John David Gilmore if he would voluntarily, with his left hand placed on the Holy Bible, and with his right hand raised, solemnly swear to, and take the following oath:

"I, John David Gilmore, do solemnly swear that I will support and defend the Constitution of the United States against all enemies, foreign

and domestic; that I will bear true faith and allegiance to the same; and that I will obey the orders of the president of the United States and the orders of the officers appointed over me, according to regulations and the Uniform Code of Military Justice. So, help me God."

Placing golden gars, which signified second lieutenant, onto the shoulders of John David Gilmore, General Drinkwater accepted a salute from John David Gilmore and commissioned him into the United States Air Force, Division of Military Intelligence. John David accepted his commission, and as a junior executive officer, reporting to General Drinkwater.

Chapter Eighteen

The General's Asset

After receiving his commission of second lieutenant into the United States Air Force's Military Intelligence Division, John David collected his luggage and personal items, and signed out of his campus housing.

Along with Major General Thorny Drinkwater, Second Lieutenant John David Gilmore was chauffeured to a Military Air Force base, where the two boarded General Drinkwater's assigned six-passenger, 310 Cessna, two engines, propeller-driven aircraft. Comfortably seated, and facing each other, with a glass of iced tea, General Drinkwater began a conversation.

"My condolences on the passing of your father, Mr. Alonzo Gilmore," General Drinkwater offered, with a tone of sincere sadness in his voice.

"Thanks, sir. As you might know General, my dad was my hero. He taught me so very much," replied Lieutenant Gilmore, while noting that he would always carry memories of his father within his heart.

"As promised, I kept up with your progress while you were in training at the Academy of Special Operations. I am so proud of your number one accomplishment in a very competitive class of seven!" exclaimed General Drinkwater, while wondering out loud, "How did you, Lieutenant Gilmore, know to change the equation to solve the last problem on the final examination?"

"Sir," Lieutenant John David began, "my father taught me to listen carefully, and to pay attention to all of what a person says. On the first day of training at the academy, and during orientation, Colonel McCleary, Campus Superintendent, noted, and I quote, that, 'To solve certain problems, a trainee may have to change the equation.' I noted, that with the packet of examination material, which included a calculator, there was an unusually shaped paper clip, which held the pages of the examination together. In reality, the unusual paper clip, was an Allen wrench. I used the wrench to gain access to the interior of the calculator, where I found the correct equation to solve the problem, which was presented in the examination," responded Lieutenant Gilmore, with a tone of calm confidence in his voice. He recited the equation. "The orbit formula, $r=(h2/\mu)/(1+e\cos\theta)$, gives the position of body m2 in its orbit around m1 as a function of the true anomaly to put into orbit a space craft."

General Drinkwater's eyes widened. "*What?* I did not know that. So, that is why you were the only trainee to solve that problem!" exclaimed General Drinkwater.

After a two-hour and fifteen-minute flight, the 310 Cessna landed safely at a top-secret Military Air Force base in the Washington, DC area. Upon deboarding the aircraft, Lieutenant Gilmore boarded a shuttle bus, which transported him to his Barracks Officer Quarters (BOQ), with orders to report to the Pentagon, in Arlington, Virginia, at 0800 hours, where he would meet with General Drinkwater to discuss in detail what his assignment would require.

Upon entering into his comfortable, fully furnished, one-bedroom plus kitchenette BOQ, John David made a telephone call to his brother, Jacob. With a full, and detailed account, John David painted a precise and verbal picture of the events and accomplishments that he, in the recent months, had experienced, minus anything top secret.

With Jacob assuring him that Blue, their pet mule, was well and in good health, John David assured his brother that he would be making a telephone call to each of their sisters and inform them on the wonderful events that had taken place in his life.

The following morning, after a restful night of relaxation and sleep, at 0800 hours, on July 2, 1963, Second Lieutenant John David Gilmore reported to General Drinkwater in his Pentagon office.

"Sir, Second Lieutenant Gilmore reporting for duty," Lieutenant Gilmore offered with a salute, and a, "Good morning, sir."

"Good morning, Lieutenant Gilmore," responded General Drinkwater, while returning the salute that was offered. He directed Lieutenant Gilmore to his private office, which was next door to his office, and noted that it was time for the two of them to get to work.

"Sir, I am prepared for work. What is it that you need for me to do?" Lieutenant Gilmore asked.

"Lieutenant Gilmore, I have three Strategic Air Command (SAC) squadrons, at three separate Air Force bases, under my command. Each is under-performing, which will adversely affect my opportunity to be promoted to lieutenant general. For me to earn a third star, each of these squadrons will have to raise their performance level. I will identify these squadrons as Projects 1, 2, and 3. Lieutenant Gilmore, I need for you to visit each squadron, evaluate their deficiencies, and provide to me a plan that will guarantee an improvement in performance and military readiness. The classified folder, which is on your desk, will provide you with the orders and pertinent information on how to carry out this assignment. I need your report, and plans for action to correct the deficiencies, no later than two calendar weeks from today's date," ordered General Drinkwater. "Lieutenant Gilmore, do you have any questions?"

"Sir, I have no questions. I will have a plan for you, and on your desk within the time frame you have ordered," Lieutenant Gilmore responded, as he saluted General Drinkwater, before asking for permission to be dismissed.

With his heart as a quiet witness, and being in the immediate presence, John David thought, *I now know why the general is famously referred to as "THORNY".*

Possessing orders that authorized him with command and control of a Lockheed F-104 Starfighter Jet, to facilitate his travel to each of the specified Air Force Base, Lieutenant Gilmore was able to complete his assignment, with recommendations, within ten working days.

Meeting with General Drinkwater, at his Pentagon office on July 12, at 1300 hours, Lieutenant Gilmore presented his findings and recommendations to his commanding officer, General Drinkwater. "General Drinkwater, sir, my report and recommendations for Project 1—problem definition is an inadequate number of aircrafts being air worthy at any given time. Of the twenty-six assigned Lockheed F-104 Starfighter Jets, eight have been grounded and deemed not air worthy for flight. These grounded aircrafts can be certified air worthy, and ready for combat by cannibalizing one Lockheed F-104 Starfighter Jet to correct the deficiencies of the grounded eight aircrafts. I recommend an aircraft mechanic team, for the task, to cannibalize the Lockheed F-104 Starfighter Jet that has been assigned to me. The mechanic team, and myself will correct the problem, and have the grounded aircrafts certified combat ready to fly within one calendar week.

"Project 2—problem definition is an unsanitary cafeteria, or mess hall. Sir, the mess hall in question is designated as the main cafeteria, and provides service to ninety-five percent of the base's military personnel. The unsanitary conditions are caused by an infestation of roaches, rats, and mice. Sir, there is just one solution to this problem—burn the cafeteria and all of its contents to prevent transporting the infestation to a new constructed cafeteria.

"Project 3—problem definition is reported to be that military wives and mothers, who represent ninety-seven percent of the shoppers at the base's commissary, and base exchange, are unhappy with the unhealthy quality and variety of products that are sold at these facilities. The military wives and mothers have formed an organization, and have forwarded their complaints to the Department of Defense. These conditions and complaints can be addressed immediately with the temporary construction of a facility that will increase the square footage and floor space of the shopping areas. This temporary solution will allow for an upgrade, and will allow us time to address the complaints of quality, variety, and quantity of products, while permanent facilities are being constructed."

"Lieutenant Gilmore, what aircraft parts will you and your aircraft mechanic team cannibalize?" General Drinkwater asked with a tone of relief in his voice.

"Sir, the aircraft parts that are needed to certify the grounded aircrafts to be combat air worthy are an air speed indicator, altimeter, AKA altitude indicator, pilot seat ejection/parachute control, and nose, and fuse lodge land gear. All of these parts will be supplied by the cannibalization of one Lockheed F-104 Starfighter Jet," John David replied with a tone of confidence in his voice. "Sir, certain problems can only be solved by changing the equation. Two plus two, in this case, can equal whatever you want it to equal," Lieutenant Gilmore concluded with a wry smile.

"Lieutenant Gilmore, do you realize the bureaucratic hoops I will have to jump through to convince the Department of Defense to approve my request for a controlled burn of a primary cafeteria on a major Air Force Base?" inquired General Drinkwater, with a tone of trepidation in his voice. "It could take months, even years to get such an approval."

"General Drinkwater, sir, I am not recommending a controlled burn, sir. I am suggesting that you will have to change the equation to solve the problem, and to change the equation, we need to execute an act of controlled arson. Sir, to succeed, with a minimum of risk for the safety of personnel, the controlled act of arson must take place when the base population is at a low. A low base population can be accomplished during the upcoming Labor Day holiday, which is the last Monday in the month of August. Sir, if you will authorize, and encourage all company commanders to issue three day passes to all nonessential personnel, that will reduce the base population to a safe level, which will reduce the risk for safety. Sir, if you will put under my command, a bomb and arson team, the problematic cafeteria will cease to exist. Sir, one additional detail—have the base fire unit department, to arrive too late to extinguish the accidental fire," Lieutenant Gilmore calmly explained, and then outlined his plan, and noted that two plus two equals whatever the general wants it to equal, and that it would clear the path for a third star, for the general.

"Lieutenant Gilmore, I approve of your recommendations and plans to resolve the issues at the base exchange and commissary. I will issue orders immediately to address those issues. As for the other two projects, I will take under advisement, and give to you what my final decision is no later than August 1, 1963," responded General Drinkwater. "And Lieutenant Gilmore, know that you are to report to the West Coast U2 Air Force Base to carry out a high altitude U2 reconnaissance mission," noted the General.

"Yes, sir, General Drinkwater. My Lockheed F-104 Starfighter Jet is being serviced for flight as we speak, and I will be departing early this afternoon." Lieutenant Gilmore saluted and asked for permission to be dismissed.

On July 28, 1963, at 0800 hours, after three days at the U2 Air Force Base on the West Coast, Lieutenant Gilmore returned to his Pentagon office to be debriefed by General Drinkwater on his mission to gather high altitude intelligence data, and the performance of the Lockheed U2 aircraft. The high-altitude, intelligence-gathering aircraft was under review by the Department of Defense for whether it should be decommissioned as an intelligence-gathering aircraft.

"Good morning, Lieutenant Gilmore," General Drinkwater said. "How was your U2 flight, and the gathering of intelligence data?" General Drinkwater received and returned a salute, and good morning from Lieutenant Gilmore.

"Sir, pilots lament, 'a safe take off, and a safe landing,' makes for a successful flight; however, the performance of the Lockheed U2—the climbing speed to reach high altitude, is too slow, and the approach for landing is too long and predictable. These characteristics make the aircraft vulnerable to attack from ground-to-air missiles. With the advancements in technology, and the launching and orbiting of satellites, I recommend a slow, but decisive decommissioning of the Lockheed U2," Lieutenant Gilmore said.

"Those, likewise, are my evaluations of the Lockheed U2 aircraft. I will be forwarding, as soon as possible, my recommendations, along with yours, to the Department of Defense that the U2 aircraft should be decommissioned as a high-altitude gatherer of military intelligence data," responded General Drinkwater. "Now let's continue on the subject of your plans and recommendations concerning Projects 1, 2, and 3. I have decided to accept your recommendations on how to resolve the existing problems on the previously identified three Air Force bases. I have identified, assembled, and authorized the aircraft mechanic team that you requested, and, also, I have authorized the arson and bomb team that you requested. As for the problems with the base commissary, and base exchange, I will be meeting with the leadership of the concerned wives and moms shoppers' organization on August 8th, at 1300 hours," General Drinkwater noted, with a confident smile on his face.

Lieutenant Gilmore said, "Me, and the aircraft mechanic team will have the grounded aircrafts air worthy and flying within a week. As for the cafeteria, along with its guests of roaches, rats, and mice, during the Labor Day weekend, will be gone, forever!" exclaimed Lieutenant Gilmore.

During the month of October of 1963, Major General Thorny Drinkwater, of the United States Air Force Military Intelligence Command, received recognition and high honors from the Department of Defense for his management and leadership skills for lifting the capabilities and morale of three major United States Air Force bases that were under his command. The Strategic Air Command (SAC) squadron was certified to be at one-hundred percent air worthy and ready for combat. The Department of Defense allocated funds for the construction of a new base commissary, and base exchange. And construction began on a new state-of-the-art cafeteria, replacing the old one, which had been destroyed by an accidental electrical fire.

On November 15, 1963, General Drinkwater, with his authority as a General Grade Officer, waived the time and grade of one year, twelve months, at the rank of second lieutenant, before being eligible for pro-motion to the rank of first lieutenant. General Drinkwater promoted, and personally pinned the Silver Bars on to the shoulders of Second Lieutenant Gilmore, which represents the rank of first lieutenant in the United States Air Force.

Along with First Lieutenant John David Gilmore's promotion, General Drinkwater recommended him to the National Aeronautics and Space Administration (NASA) for astronaut training.

On December 1, 1963, General Drinkwater was tasked with the planning of extracting an American asset from the island nation of Cuba. With this assigned responsibility, General Drinkwater called upon the skills of First Lieutenant Gilmore to study the intelligence reports surrounding this American asset, and devise a plan of action for a safe extraction. On December 3, 1963, at 0800 hours, First Lieutenant Gilmore met with General Drinkwater, and presented a detailed plan for an extraction.

First Lieutenant Gilmore, possessing two identical folders marked TOP SECRET, with color-coded tabs on specific pages, presented one folder to the General, seated himself facing his commanding officer, and began the briefing. "Sir, to extract the said American asset, I will have to be assigned to the United States Embassy, which is located in the city of Havana, Cuba. My assignment and job description must be as an English to Spanish translator. As you may recall, I speak fluent Spanish. I estimate the time to acclimate myself, and blend into the culture and local population of the community, will take nine to twelve months before we can safely extract the American asset. The method of extraction will be by a naval water craft. My plan will re-quire me and the American asset to utilize SCUBA gear. We will then swim three nautical miles to international water, to be intercepted, and safely transported to a safe beach.

"In the meantime, while assimilating in the local population, my-self and the American asset will establish a routine of swimming at a specified location at different times of the day. On the predetermined time and date of extraction, sir, the SCUBA gear, along with locator frequency, and signal, for the two of us, must be at the predetermined set of coordinates," Lieutenant Gilmore said, with a tone of profes-sional perfection in his voice.

General Drinkwater smiled. "Lieutenant Gilmore, I am always im-pressed with the simplicity, and vital details contained within the plans that you present to me; and may I add, the plans have all worked. I will approve your plans and recommendations, and present them for approval to the senior chain of command." Pleased with the plans, General Drinkwater confidently ordered Lieutenant Gilmore to pre-pare to be assigned to the United States Embassy in Havana, Cuba as an English to Spanish interpreter.

"Sir, thanks for the confidence that you have in me," Lieutenant Gilmore said. He saluted and requested permission to be dismissed.

Returning to his BOQ living quarters, and while packing his lug-gage to travel to his new assignment, John David made a telephone call to each of his siblings, and with sadness and a solemn moment of reminiscing, the siblings wished each other a Merry Christmas and a Happy New Year, and Jacob noted that their pet mule, Blue, was well.

Following the telephone conversation, and pondering in his heart, if he would ever return home for a family reunion, John David put on headsets to, once again, hear the recorded voices of his siblings that were made during the funeral of their father, Alonzo Gilmore.

On January 15, 1964, First Lieutenant John David Gilmore arrived at the Havana, Cuba International Airport, and was transported by United States Embassy personnel to his living quarters, which resembled the one bedroom, efficiency kitchenette, that he was assigned to live in during his assignment at the Pentagon.

Thursday, November 12, 1964, at approximately 1400 hours, after days, weeks, and months of methodically and patiently establishing a routine to assimilate in to the community, with the American asset, Lieutenant Gilmore, along with his asset, entered the chilly water off the beach of Havana, Cuba, and perfectly executed the approved plan that he had presented to General Drinkwater, to extract an American asset.

As the evening sun began to set over the Atlantic Ocean, on November 12, a United States Navy vessel safely landed at a Florida beach, carrying Lieutenant Gilmore and a high-profile American asset, where they were welcomed ashore by a pleased and happy Major General Thorny Drinkwater.

Chapter Nineteen

I Am Beginning To Say Farewell

"...and now my vision and fear of not returning to my country, which I loved, was made clear. I look back once more – for the last time – before devising a plan to remain in Vietnam with my family. I see myself with my adopted family by my side, journeying along the road of life." Words of Major John David Gilmore

On January 15, 1965, after completing several difficult and high-profile assignments, with an Officer Evaluation Review (OER) rating score, that had been recorded in the .05 percent percentile of the general grade officers of all US Military branches of service, the president of the United States of America, and commander in chief of the US Military, signed orders promoting Major General Thornton "Thorny" Drinkwater, to the rank of lieutenant general of the US Air Force.

At the promotion ceremony, which was attended by the Secretary of Defense, along with senior members of his staff, the newly promoted Lieutenant General Drinkwater, called on First Lieutenant

John David Gilmore to pin the three stars on to his shoulders. Once the insignia of three silver stars had been pinned on to his shoulders, Lieutenant General Drinkwater removed from the inside pocket of his dress blue jacket, a pair of double-silver bars, referred to as rail road tracks, which is the insignia for the rank of captain of the US Air Force.

Informing the audience that, with the authority of a lieutenant general of the US Air Force, he had waived the time and grade of eighteen months of a first lieutenant, before being promoted to the rank of captain, and therefore, was promoting First Lieutenant John David Gilmore, to Captain Gilmore of the US Air Force, and he pinned the double silver bars onto the shoulders of Captain Gilmore.

After a salute from each of the proud and newly promoted officers, and a loud applause from those in attendance, who witnessed the double promotion ceremony, Lieutenant General Drinkwater, with a salute to the Secretary of Defense, requested permission to be dismissed.

After returning to their Pentagon offices, and enjoying the moment and events surrounding their promotions, General Drinkwater informed Captain Gilmore that he would have to make a decision on which he would choose for his career.

"Captain Gilmore, with my promotion, I have received assignment to U-Tapao Air Force Base, Thailand, to command the Fighter Squadron that give support to the B-52 bombers. I have, also, received the authority to assign you to the National Aeronautical Space Administration (NASA) for astronaut training. The decision is for you to make. If you elect, now, to be assigned with me, you can later on in your career, accept the assignment to NASA," General Drinkwater advised Captain Gilmore, and then he noted that, "Captain Gilmore, you will have my total support, regardless of your decision."

"General Drinkwater, sir, where ever you go, sir, I will go; where ever you lead, sir, I will follow!" exclaimed Captain Gilmore. "When are we to report for our new assignment?"

"Son, uh, I mean Captain Gilmore, we are to report to the commander of the U-Tapao Air Base, Thailand on or about February 2, 1965, 1300 hours, local time. We are to begin with our out processing today."

Both officers saluted each other, while noting in their hearts that their relationship was growing toward what a father and son would share.

"Permission to be dismissed, sir," Captain Gilmore requested.

"Permission granted."

On February 1, 1965, at approximately 0600 hours, General Drinkwater and Captain Gilmore, along with fifty-seven officers and enlisted military personnel, and a crew of seven, boarded a Boeing KC-135 Stratotanker, which had been converted to a military troop passenger plane. They landed at U-Tapao Air Base after a twenty-two-hour flight.

Transported by base shuttle bus services to their adjoining BOQ housing, which had been specially ordered by General Drinkwater to allow for company-grade and general officers to reside in the same building, General Drinkwater and Captain Gilmore signed into their separate, comfortable, one-bedroom living quarters.

After a shower, a shave, and dressing in fresh combat uniforms, at 1300 hours, General Drinkwater and Captain Gilmore reported to the commanding four-star general of the U-Tapao Air Base, for the details of their official assignments and duties, which included, but were not limited to:

- Maintain and keep the B-52 Bomber Squadron at 95-100 percent worthy for flight.

- Maintain and keep the Lockheed F-104 Fighter Jet Squadron at 95-100 percent worthy for flight.

- Establish, train, and certify pilots for an AH-1 Helicopter Cobra Fighter Platoon.

- Establish and train an elite sniper team.

- Maintain at 100 percent air readiness, the F-104 aircraft, for both low- and high-altitude flight intelligence data gathering.

- Maintain high morale among all military personnel with timely Rest and Recuperation (R&R) to secure locations.

For the American Military to be successful in their efforts to win the Vietnam War against the communist-led forces of the Viet Cong, it was strategically imperative for the B-52 Bomber Squadron, and the support forces of the Lockheed F-104 Fighter Jet Platoon, at the U-Tapao Air Base to perform at 95-100 percent efficiency. This responsibility was left to the leadership skills and command of General Drinkwater and Captain Gilmore.

With orders, and backing from General Drinkwater, who gave to Captain Gilmore support personnel of his choosing, budget, and financial control, and the authority to put in place all of the aforementioned assignments that had been ordered by the commanding general of the U-Tapao Air Base, Thailand, General Drinkwater and Captain Gilmore completed their assignments with flying colors.

Beginning with the first week of March, and throughout the last week of October, 1965, while working twelve to sixteen hours per day, including Saturdays and Sundays, Captain Gilmore, along with his support personnel, was successful in accomplishing the assignments that he was charged to deliver.

In addition to overseeing the upgrade of the B-52 Bomber Squadron at the U-Tapao Air Base to the required 95-100 percent flight worthiness in order to successfully support the American Military forces in their fight against the communist led Viet Cong Army, Captain Gilmore was also required to fly F-104 Starfighter Jet low-altitude reconnaissance missions to collect military intelligence data order.

On November 17, 1965, just as the sun was rising over the Cambodian-Vietnam border, Captain Gilmore's Lockheed F-104 Starfighter Jet was struck by a ground-to-air missile. Losing control of the aircraft, and while giving out a Mayday emergency call, Captain Gilmore initiated his emergency seat parachute, while triggering an explosive device that destroyed the top-secret aircraft along with its intelligence-gathering equipment.

Except for a few scratches and minor bruises, Captain Gilmore successfully landed into a muddy rice paddy near a friendly Vietnamese Hamlet, which was approximately fifteen miles southeast of the Cu Chi, US Army Military Base, home to the 25th Aviation Helicopter Squadron. Immediately upon his landing into the muddy rice paddy, Captain Gilmore was rescued by a young Vietnamese woman, and two pre-teen boys, who were her brothers.

The female, who spoke perfect English, introduced herself to be Hao Hong Nguyen, and her brothers, Duc, the oldest, and Dung. After securing his parachute, which contained a morse code communicator device, Captain Gilmore sent a morse code message to General Drinkwater, informing him of his shoot down, and the geographical coordinates of his location, and that he was well, and in the hands of a friendly Vietnamese family.

Welcomed into the home of the Nguyen family, the sister, Hao Hong, provided Captain Gilmore a robe to wear, while she washed his muddy flight suit, and all of the other clothing that needed washing. Keeping on his underwear, Captain Gilmore offered his military fatigues, which he was wearing in addition to his flight suit, to be washed.

Waiting for his clothes to dry by the sunlight, and outdoor gentle breeze, along with an out-door temperature of approximately 75 degrees Fahrenheit, Captain Gilmore was served a meal consisting of broiled fish, rice, and sweet tea, served at room temperature.

With the sun light, and gentle warm breeze, Captain Gilmore's flight suit and camouflaged fatigues were dry enough for him to dress into, within a couple of hours.

Anticipating to be extracted by a rescue team within a few hours and while sitting at the dining room table and sipping room-temperature sweet tea, a conversation between Captain Gilmore and Hao Hong ensued.

For the first time in his twenty-four years of life, John David felt an intense physical attraction for a female. Hao Hong, wore a drab, loose-fitting, black, pajama-like jump suit. She stood approximately 5 feet 2 inches in height, and weighed approximately 105 pounds. Hao Hong wore her coal-black hair, in a boyish short cut, having the same wavy texture of John David's mother, Rose. Her skin coloring was light brown, and her eyes were a captivating sky-blue in color, framed with a perfectly sculptured nose, with pouting lips, and ivory-white teeth, and a gentle smile. John David, collecting his emotions, which was born out of his attraction to Hao Hong, managed to verbalize, "Ma'am, you speak fluent English. Who taught you to speak English, so perfectly?"

Hao Hong directed her brothers to go to their room, and study their lessons, as she was home schooling them. Hao Hong, who was physically attracted to her rescued guest, took a deep breath to calm her nervous emotions and managed to reply, "My mother was French, and an English to French language translator for the US Embassy in Saigon. She was highly educated, and home schooled me, and my brothers, and taught us her native language, French, and English. My father was Vietnamese, and highly educated, so we learned to speak Vietnamese during the course of everyday life. When I successfully completed the equivalency of high school, and was certified to attend college, I received a scholarship to attend the Language Institute at George Washington University, in Washington, DC, in America, where I concentrated on English. I completed my studies in America in June of 1962, and became a translator of Vietnamese, French, and English, and as my mother before me, became a translator at the US Embassy in Saigon. I worked there until October of 1964, that is when my mother died. My father was killed in 1962, while fighting the Viet Cong; it was the year when I completed my studies in America. So, when my mother lost her battle with cancer and died, I had to return home to care for my young brothers," exclaimed Hao Hong. With her arms folded over her chest and pacing around the small dining room, she confided, "I have never before known an American with the physical characteristics that you have, sir," she said with a nervous tone in her voice.

"I am the offspring of mixed-raced parents. The mother of my father was a negro, and his father was a white man. The father of my mother was white, and her mother was Creole, which is a mixture of French and negro," John David explained, while gazing into the eyes of Hao Hong, and pondering in his heart if their meeting was the design of divine intervention to cause him to fall into Hao Hong's rice paddy.

Hao Hong said, "Captain Gilmore, do you believe in—"

A Special Forces rescue team arrived at the home of Hao Hong to extract Captain Gilmore.

With just a few minutes to say goodbye, and while placing a business-like card into the small, work-calloused hand of his rescuer, with information of how to contact him, Captain Gilmore promised Hao Hong that he would see her again. He entered into the rescue helicopter, and waved goodbye.

Transported to the US Embassy in Saigon, he was greeted by General Drinkwater, who insisted that Captain Gilmore be examined by the embassy's medical staff.

After receiving a thorough medical and physical examination by the embassy's medical staff, Captain Gilmore was found to have not suffered any injuries, other than scrapes and bruises, and was released to resume his duties.

Aboard General Drinkwater's assigned Cessna-310 aircraft, and while enroute to the U-Tapao Thailand Air Base, Captain Gilmore gave a detailed briefing of his reconnaissance data-gathering flight, and the subsequent shoot-down. He included his meeting, and his intense physical attraction, with Hao Hong Nguyen, his beautiful rescuer.

"General Drinkwater, sir," Captain Gilmore said, "I have never been in a relationship with a woman. I am a virgin, but I want to get to know Hao Hong. Sir, will you order a security investigation of her? And if she passes the vetting, I would like for your permission, and approval, to spend some time with her."

General Drinkwater suppressed memories of himself choosing a career over love, and leaving behind, in South Korea, the young woman and love of his life, during the Korean War, when he, too, was a captain. "Son—sorry—I mean Captain Gilmore; I will submit the necessary paper work to execute a top-secret investigation of Ms.

Nguyen. From what you have shared with me, concerning her, the investigation should not take that long," responded General Drinkwater, then advised John David to always follow his heart.

On November 19, 1965, two days removed from the shoot down of his Lockheed F-104 Starfighter, Captain Gilmore was summoned by GeneralDrinkwater to attend a top secret briefing at the US Embassy in Saigon to discuss and review a recent rash of assassinations of US Intelligence officials, and their assets in the city of Saigon, the capital of Vietnam.

The top-secret briefing was adjoined with General Drinkwater and Captain Gilmore receiving top secret data for their eyes only that included autopsies, photographs, times of day, and locations of the assassinations, and reports from local pedestrians who were at the scene of the assassinations.

With a salute, and request for permission to be dismissed, General Drinkwater assured the commanding officer, who held the rank of colonel of the United States Air Force Intelligence Division, that after analyzing the data, a report and plan for action would be on his desk within three weeks.

On December 3, 1965, at 0800 hours, Captain Gilmore, after meticulously analyzing and evaluating the information and intelligence data of the assassinations that had occurred in Saigon, met in a secure briefing room at the office of General Drinkwater, where the two officers discussed the intelligence data, and recommended plans of action to neutralize the assassin operating in Saigon.

"General Drinkwater, sir," began Captain Gilmore with a customary salute. "Good morning. Sir, I have completed my analysis of the intelligence data that was provided to us." He handed to General Drinkwater a top-secret folder. "Permission to begin with the briefing, sir?"

"Captain Gilmore, permission to begin with briefing, granted," responded General Drinkwater, with a return salute, and a good morning.

"Sir, my analysis reveals the following. One: The assumption that there is a single assassin is incorrect. There are a minimum of two assassins, perhaps three, working as a team. This conclusion is based on the facts that on September 15, 1965, within a time frame of two minutes, a distance of one land mile, as the crow flies, two assets were assassinated. In addition to these facts, ballistics confirms that two, .22 caliber weapons were used, establishing the irrefutable fact that two assassins were responsible. Two: Saigon intelligence analysts have determined the assassin to be male. Sir, my analysis, and I am 99.9 percent certain, is that the assassins are female. Research of established data has determined that the choice of weapon for a female is a small caliber pistol, this is due to hand size and strength of the assassin."

General Drinkwater silently nodded his head.

"Three: The autopsy reports establish the entry wounds to have an upward projection when the projectile entered the heart, and that the shooter was at a very close distance. These findings are based on the facts that the victims were 6 feet to 6 feet 2 inches in height. The upward projection of the projectile, establishes the assassin to be at a maximum of 5 feet 1 inch in height, causing the entry wound from a very close distance, and having an upward projection. And four: My analysis directs me to believe that the assassins are female, and they are disguising themselves in maternity clothing, which allows for them to enter into very close distances of their targets. These findings are based on black and white still-photographs that were secured from pedestrians who were in the area during the approximate time of the assassinations. In each photograph is the image of a female wearing clothing that a pregnant woman would be wearing.

"Sir, my recommendation, to neutralize the assassins, is to identify and secure the tallest building within the fifteen-block perimeter of where the assassinations have occurred, then a sniper nest must be constructed on the roof-top, which will be monitored twenty-four hours a day, and seven days a week. Also, we must utilize on-the-ground, foot intelligence officers to identify and photograph all females who are dressed in maternity clothing. Sir, for this assignment, I recommend the sniper team which I have personally trained," concluded Captain Gilmore.

"And sir, the assassins were in possession of a boat load of confidential information on our intelligence officers. My gut is telling me, that it is possible, that a mole intelligence officer is in the ranks of the US Air Force Intelligence Division," Captain Gilmore emphasized.

General Drinkwater said, "Captain Gilmore, I accept your report and recommendations to neutralize the assassins, and will issue orders for immediate implementation." General Drinkwater, again, was impressed with the detailed analysis of the report.

"Oh, I almost forgot. The investigation of Hao Hong has been completed, and she passed with flying colors," General Drinkwater said while advising Captain Gilmore to follow his heart, and with a salute, gave permission to be dismissed.

Back at his BOQ living quarters, Captain Gilmore had an envelope in his mail box that was addressed from Hao Hong.

The letter inside of the envelope, read:

November 28, 1965

Hao Hong Nguyen
CU CHI Hamlet #19

Dear Captain Gilmore:

I am writing to wish that all is well with you and that you have recovered from your scratches and bruises.

The day that you fell from the sky, I was praying to God that he
would send to me a companion. I have never been with a man and being mixed race, we both know how difficult it is to meet a companion.

I pray that this letter will get to you and we can meet again.

May God answer my prayers.

Always,
Hao Hong

Chapter Twenty

Now Acceptance: From Vietnam -- 'Farewell' !

The following day, while working with General Drinkwater on finalizing the plans, and

selecting the personnel for the sniper team, and equipment to neutralize the assassins who were operating in Saigon, Captain Gilmore shared the contents of the letter that he had received from Hao Hong, and requested counseling from General Drinkwater on what he should do.

Responding in the tone of a caring father, and not wanting John David to make the mistake that he had made when he chose career over the woman he loved, General Drinkwater, with a commanding tone in his voice, prepared Captain Gilmore to accept what he was going to do as his senior officer, and that he wished for him to accept, and obey his recommendations, as if they were orders. "I am going to assemble, and order the Special Forces team, who extracted, and transported you from Hao Hong's hamlet, to escort you to return to her home, and confess to her your attraction, and feelings for her. While there, I will order the Special Forces team to install a secure, top secret, military-grade communication system in her home, so that you and Hao Hong can communicate with each other on a regular and set schedule. And third, I suggest for you to take a rest and recuperation vacation with Hao Hong, the woman whom you are apparently in love with, to take a week of R&R to a location that the two of you choose. My records show that you have never taken a vacation since being on active duty. So, Captain, as your commanding officer, I am going to order for you to take some time off, and enjoy some R&R, and I want for you to prepare to depart with the Special Forces team as soon as possible, and to inform me where you and Hao Hong will be vacationing," the General ordered. "On returning, you will be assigned to neutralize the Saigon assassins, which will require nine to twelve months of disciplined solitude, and is the equivalency of watching grass grow," General Drinkwater concluded with a chuckle.

On December 19, 1965, at approximately 0900 hours, General Drinkwater, after providing Captain Gilmore with his personal, no-limit military credit card, personally planned the logistics for Captain Gilmore to travel to the home of Hao Hong, and her two brothers, in the Cu Chi Hamlet of South Vietnam, which was approximately ninety miles south of Saigon. Captain Gilmore, escorted by a Special Forces team, landed a UH-1 Helicopter onto the front yard of Hao Hong's home.

Surprised! Happy! Overjoyed! Words could never explain her emotions as Hao Hong ran to meet John David. She was able to verbalize through a stream of happy tears. "I prayed for you to come back to me."

"And I knew that you would be waiting," John David exclaimed as they embraced each other. The two entered into the sparsely furnished, but comfortable, hut that Hao Hong and her brothers called home.

While sitting at the dining room table, John David removed the letter that Hao Hong had written to him, and explained to her, "Based on the content of your letter, my commanding officer, General Drinkwater, had ordered that I take a vacation of rest and recuperation, and invite you to go with me. Hao Hong, will you take a vacation of R&R with me?" John David asked. "Also, I too, am a virgin, and have never been in a relationship with a woman." He felt his cheeks swell as he shifted on his feet.

"Before I answer, I want to know if you believe there is a God. I do believe there is a God? I am a Christian, and every day I prayed to God that he would send to me a companion, and you fell from the sky. I didn't expect that part." The two filled the conversation with laughter. "And furthermore, what will I do with my young brothers?" asked Hao Hong with a tone of faithful reservation in her voice.

"Yes, I do believe there is a living God, and that Jesus Christ is His only begotten Son, and that the Holy Spirit, rest rule and abide within all believers. And, I, too, prayed for God to give to me the woman who was in my mind's eye, and Hao Hong, you are she! And if you will consent to go with me, we will take your brothers with us," John David promised.

His communication team, in the background, were giving approving high-fives for their captain's courage to ask Hao Hong to go on R&R with him. "I want y'all, right now to get busy, and install the secure, military-grade communication system into the home of Hao Hong," Captain Gilmore ordered with a sheepish grin on his face.

"We know," replied the communication's team leader, while noting that the General had explained the need for his captain and Hao Hong, to have the capabilities to communicate on a pre-planned schedule.

"Yes, I, we, and my brothers will go with you on R&R! Can we, please, go to Orleans France? My mother's sister, Aunt Camille, lives there. It has been years since we last saw her," Hao Hong happily responded, and she instructed her brothers, Duc and Dung, to pack their luggage and to notify the chief of their Cu Chi Hamlet of their vacation plans, and she wished for the other twenty-three families to take care of their garden, along with the chickens and ducks, as they were going to visit their Aunt Camille, who lived in France.

At approximately 2000 hours on December 19, 1965, Captain John David Gilmore, Hao Hong, and her brothers, Duc and Dung, boarded a US Military KC-135 aircraft, which had been converted for passenger accommodation and travel, at the Tan Son Nhat International Airport. The flight passenger manifest, which consisted of fifty-three passengers, plus a crew of five, safely landed at the Orly US Military Air Force Base, France, after a flight time of approximately fifteen hours.

After claiming their luggage, Captain Gilmore and his newly adopted family were met at the US Air Force car pool, by an Air Force guide, who issued to Captain Gilmore a set of car keys to a dark blue, 1964 four-door, five passenger Chevrolet automobile, and was provided with instructions, and a map of the base, with outlined directions on how to locate their housing.

Arriving at their comfortable three bedroom, with two bathrooms, complete with a shower, a tub, and a kitchen, which had been stocked with an assortment of breakfast and dinner food, including a small turkey breast, and a two-pound boneless ham, all topped off with a twenty-four-can case of Coca Cola.

The blended family decided to take a shower, rest, and sleep, before traveling on the following day to the city of Orleans, France to visit with the siblings' Aunt Camille and Uncle Jules DuBois.

Waking early as planned, and after a family effort of preparing and enjoying a filling and nutritious breakfast of sausages, eggs, and hot cereal, along with a choice of milk, juice, and coffee, John David and his adopted family began their drive to Orleans, France. During the three-hour drive, which was filled with casual chit chat, the blended family began to better know each other, as they shared their dreams and hopes for their futures.

Speaking fluent English, and with a warm introduction and greeting, the aunt and uncle of Hao Hong, and brothers, Duc and Dung, welcomed Captain Gilmore and his adopted family into their home. The two story, sand-stone brick home sat on a fifty-acre farm, where the DuBois family grew grapes and produced a variety of wines.

The four-hour reunion allowed the niece and nephews, and their Aunt Camille and Uncle Jules, to reminisce and discuss the events that had occurred during the intervening years since their last visit.

After a few boring minutes of listening to their sister, aunt, and uncle talk about times past, Duc and Dung requested, and were given permission, to go outside to explore the farm and its vineyards while their sister and aunt, in the process of preparing dinner, discussed the miracle of how John David had fallen into her life.

In the meantime, Jules was giving John David a tour of the grape farm and wine production business that he and his wife, Camille, owned and managed.

John David shared with Jules how he and Hao Hong had met, and that he planned to ask her to marry him.

Listening intently, and with amazement, as John David described the day and miracle of divine intervention of meeting Hao Hong, Jules responded with an amazed look on his face, and excitement in the tone of his voice. "Yes! You must marry her. God put you and Hao Hong together!"

At the dinner table, and meal, which included baked chicken, mashed potatoes, green beans, and home-made rolls, along with a glass of Chenin Blanc wine for the adults, and sweet tea for Hao Hong's brothers as Jules proudly noted that the seven-year-old bottle of wine had been produced by the DuBois's Winery. While holding hands with each other, Jules offered a prayer of grace and thanks for the food, and prayed for God to bless John David Gilmore, Hao Hong, Duc, Dung, and his precious wife, Camille. "Amen."

After dinner, John David's adopted family, and aunt and uncle, embraced each other, and said goodbye, with the promise of not to take as long from the last visit to the next.

On the return drive to their Orly Air Base living quarters, as her young brothers had fallen asleep in the back seat, Hao Hong used the convenience of the design of the bench front seat, and chose to sit close to John David as he drove. She held his right hand and desired for him to make love with her.

On the third day of their vacation, Hao Hong devised a plan for her brothers to attend an evening movie at the base's theater. At 1800 hours, John David and Hao Hong drove her brothers to the movie theater and combination arcade center, with a financial allowance for the two brothers to attend a movie, visit the local McDonald's restaurant, and play some video games, and to be prepared to be picked-up at 2200 hours at the same location where they were dropped off.

Back at their living quarters, and alone together for the first time, words were not needed to describe their physical, and sexual, attraction for each other. They embraced with opened mouths and kisses, which left each of them breathless, but, yet, finding the energy to whisper in unison, "I'm in love with you."

After tearing themselves apart and undressing themselves, they moved toward their respective bathroom to take a shower. Hao Hong, naked, entered into the bedroom of John David and gave her 105-pound, pint-sized, but perfectly sculptured body and hardened breasts and nipples, to the waiting mouth of John David, as he laid naked and mercilessly aroused, with a throbbing erection.

While both John David and Hao Hong were virgins, their love making was divine, deliberate, and totally draining.

Hao Hong passionately kissed the open mouth of John David, while quietly murmuring, "John David, I am in love with you. I will always be faithful, and belong only to you. Please, don't just use me, and then leave me."

While this was the first woman for John David to make love with, he vowed to God that one day Hao Hong would be his wife, and until his death; therefore, as he knew her, and their love making was patient, urgent, repetitive, and exhausting, it was undeniably *pure*. "I will never leave you! I am not using you. And I am going to marry you," John David vowed, and he smothered Hao Hong's open mouth with passionate kisses.

Because their sister and John David arrived thirty minutes late to meet them at the prearranged location for the ride to their living quarters, Duc and Dung were curiously suspicious of their lateness.

The following day was December 24th, Christmas Eve, and John David had woken at 0600 hours. For himself and his adopted family, he prepared breakfast, which consisted of scrambled eggs, bacon, sausage, and pancakes, accompanied with milk, orange juice, and coffee.

At the breakfast table, and after greetings of good morning, with a prayer to bless the food, John David presented to the family, "Who wants to spend the day touring the city of Paris, and shop for Christmas gifts?"

The suggestion was unanimously approved with a loud and joyous, "Yes!"

The day of touring, sightseeing, and shopping was enjoyable, and awakened memories in the mind of John David of when his mother and father would take him and his sisters and brothers to the big cities of Cairo, Illinois, and Cape Girardeau, Missouri to do their Christmas shopping for gifts and toys.

With instructions to purchase a gift for each family member, and that everyone would meet at a designated cashier for their purchase to be paid for with the credit card provided by General Drinkwater, the blended family set out to do their Christmas shopping.

Back at their living quarters, while exhausted yet gleefully decorating the artificial Christmas tree they had agreed upon to purchase, Hao Hong and her brothers began to speak and reminisce in their native language of Vietnamese of the Christmas holiday seasons before the deaths of their father and mother.

In the meantime, John David occupied himself in the kitchen marinating the turkey and preparing the ham for Christmas dinner.

After midnight, believing her brothers, exhausted from the day's activities, were in a deep sleep, Hao Hong took a warm shower and slipped into a revealing black negligee, which she had purchased at a Paris lingerie shop. While wearing a full-length top coat to cover the sheer, black negligee, she quietly entered into John David's bedroom.

Removing the top coat, and with a slither of moon light seeping through the cracks of the bedroom's window curtains, the silhouette of Hao Hong's negligee-covered body aroused John David to an immediate erection. He caressed her wetness against his waiting man-hood.

"I am a little sore from our love making from yesterday, but I need for you to make love to me, again!" Hao Hong begged, as she guided John David's throbbing erection into her waiting wetness. "Gently," she seductively whispered in his ear.

After two hours of quietly making love with John David, Hao Hong dressed herself in only the top coat, placed the negligee into a coat pocket, and slipped back to her bedroom, where she fell into a deep and relaxing sleep.

The activities of Christmas Eve were cause for everyone to have a late sleep in. Finally, awake at approximately 0900 hours, the blended family excitedly began to open their Christmas gifts.

- From John David to my love Hao Hong: A gold wedding ring.

- From John David to brothers, Duc and Dung: A chess board, with chess pieces.

- From Hao Hong to my love John David: A journal.

- From Hao Hong to my brothers, Duc and Dung: Books, and an assortment of school supplies.

- From brothers Duc and Dung to our sister, Hao Hong: A bottle of Chanel perfume.

- From brothers Duc and Dung to John David, our future brother-in-law: A bottle of Old Spice Shaving lotion.

- From Duc to my brother, Dung: A race car model set, to be assembled.

- From Dung to my brother, Duc: A model air plane set, to be assembled.

While Hao Hong was in the kitchen preparing and cooking the Christmas dinner of turkey and ham, with side dishes of dressing, sweet potatoes, cranberry sauce, and rolls, John David was busy in the living room teaching his future brothers-in-law how to play the game of chess, and getting to know each of them personally.

After teaching the pre-teen brothers the name, position on the chess board, and the significance of each chess piece, John David asked, "Which of you two will take the first move, with a white chess piece?"

"I will," responded Duc. "I am the oldest, and I will turn thirteen on January 31, 1966."

"And I will turn twelve on March 15, 1966," Dung noted, with a smile of concession.

After several hours of John David teaching Duc and Dung the nuances and strategies of the game of chess, Hao Hong made the announcement that the Christmas dinner was ready, and for everyone to wash their hands before coming to the dinner table.

Seated in the head chair, with Hao Hong sitting at his right, John David encouraged each member of the blended family, beginning with the youngest, Dung, to offer a prayer of thanks to God for his Grace, and Blessings.

The prayer of Dung was, "Father God, and Mother Mary, please accept my thanks for a blessed Christmas, and for giving to us Captain John David! And dear God, allow our mother and father to rest in peace. Amen!"

The prayer of Duc was, "Precious God, the Father of our Lord, and Savior Jesus Christ, please accept my thanks for a blessed and merry Christmas, and for giving to me and my brother a father figure in the person of Captain Gilmore. And dear God, please continue to bless the dreams and hopes for our sister, Hao Hong."

Hao Hong said, "Father God, and the guiding hand of Mother Mary, and the eternal presence of the Holy Spirit, who rest, rule, and abide with in me. Thanks for protecting me from all hurt, and danger, and for giving me strength and wisdom to lead and guide my brothers. And, dear Jesus, thanks for putting John David in to my life! Amen."

The prayer of John David was, "Eternal and Merciful God, you have answered all of my prayers, dreams, and hopes when you put Hao Hong and her brothers into my life. I pray that you will give to me as my wife, in holy matrimony, Hao Hong. We ask you Father God to bless the food, and bless our bodies to receive this daily bread, for health, healing, and strength."

While joining hands, the blended family gave a joyous, "Amen!"

At that moment, John David was shown a vision that he would never leave Vietnam.

The Christmas dinner was served with a glass of wine for John David and Hao Hong, which had been gifted from the DuBois family, while Duc and Dung were served their favorite beverage, Coca Cola.

"Who wants a serving of cherry pie, with a scoop of vanilla ice cream?" Hao Hong inquired as she removed the dinner plates from the table.

"I do," responded Duc, with his brother nodding his head, with a yes.

"Don't forget me," John David said, as he reminded his adopted family that they would be departing for Vietnam at 0900 hours, on tomorrow, December 26, 1965.

"And I will be having dessert, too," Hao Hong said quietly, as she began to serve.

Awoken early, and just as the morning sun was rising on a cold, French December morning, John David, along with brothers Duc and Dung, assumed the responsibilities to clean their living quarters, while Hao Hong prepared for their flight to Saigon, sandwiches from the left-over ham and turkey, which they'd enjoyed from their Christmas dinner, along with several cans of Coca Cola.

At 0900 hours, two airmen arrived at the living quarters of Captain Gilmore and his adopted family. One airman took possession of the motor pool vehicle, which had been used for transportation by the blended family. The second airman, chauffeuring a seven-passenger, blue Chevrolet van, transported the blended family to the Orly Air Force Base for their flight to Tan Son Nhat International Airport, Saigon, Vietnam.

Aboard the KC-135 military passenger plane, with a total of fifty-seven passengers, and five crew members, John David and Hao Hong sat side-by-side, and holding hands, while Duc and Dung, seated in adjoining seats across the aisle way, were absorbed in competitive games of chess, while munching on turkey and ham sandwiches, and drinking Coca Cola.

During the entire fifteen-hour flight, which had departed at approximately 1300 hours local time, John David and Hao Hong, except for an occasional visit to the bathroom, held each other's hands while expressing their love and commitment for each other, and plans for their future together.

After arriving safely at the Saigon Airport, Captain Gilmore, along with his adopted family, was met by the same Special Forces team who had escorted them for their departure. The team was now charged with the mission to escort Hao Hong and her brothers to their home in the Cu Chi Hamlet.

Safely in the confines of their home, Captain Gilmore gathered Hao Hong and her brothers together for the purpose of teaching each of them how to use the communication system that had been installed before they left to take their vacation.

Once Captain Gilmore was confident that his future wife and her brothers were trained sufficiently on the safe and secure usage of the communication system, Hao Hong, in the presence of Duc and Dung, embraced and kissed him passionately.

The two brothers vowed to their future brother-in-law, that they would protect their sister, his future wife.

The Special Forces team safely returned Captain Gilmore to his BOQ at the U-Tapao Air Base, Thailand. On December 28, 1965, at 0800 hours, Captain Gilmore, fresh from his vacation of R&R, reported to General Drinkwater for duty.

"Good morning, sir. Captain Gilmore reporting for duty," John David said with a salute. "Thanks for the use of your credit card, and the VIP plans, which you so kindly made for me and my adopted family. I will take responsibility for the credit card bill," John David said with a tone of gratitude in his voice.

"At ease, Captain," responded General Drinkwater followed by a return salute and a good morning. "You're welcome, and the use of my credit card was my Christmas gift to you, Captain Gilmore. However, there is some bad news. There has been another assassination," the General said. "Your sniper team was able to collect a trove of intelligence data, along with some photographs of the suspected assassins. The top-secret folder is on your desk for analysis, and a plan for action," concluded General Drinkwater.

"Sir, I will review and analyze the intelligence data, and report back to you within three days, which is January 1, 1966, the first day of the new year," responded Captain Gilmore before saluting, and requesting for permission to be dismissed.

At six o'clock, local time, in Wyatt, Missouri, on December 31, 1965—New Year's Eve—John David and his sisters Martha, Sarah, and Rachel, along with brother Jacob, via a well-planned conference telephone call, had a joyous family conversation.

After thanking God for watching over and keeping each family member well, including their pet mule, Blue, and that their dreams were being fulfilled, John David added, with details, that he had met his future wife, Hao Hong, which provided him with the honor to receive their mother's wedding ring for his bride, because he was the first of the siblings to be engaged for holy matrimony.

John David received prayers and blessings from each of his siblings.

Jacob assured his younger brother that he would mail their mother's wedding ring as soon as possible and advised John David that he must now record his wife to be his beneficiary, and that the farm was now financially successful.

With a tone of sadness in each of the siblings' voices, and while pondering in their hearts that they would probably never see their brother again, and remembering their beloved father and mother, along with brother Jonathan and sister Mary, all who were resting in heaven, the Gilmore sisters and brothers, in unison, said, "Goodbye, and *Happy New Year!*"

At 0800 hours, on New Year's Day, 1966, Captain Gilmore met in the secure briefing room of General Drinkwater's office to present his analysis and recommendations for the resolution to the assassinations of intelligence personnel in the city of Saigon.

"General Drinkwater, sir," began Captain Gilmore with a salute, "my forensic analysis of the intelligence data of the recent assassination, reveals several important notes. One: The .22 caliber weapon was the same that was used in five of the previous assassinations. Two: The Method of Operation (MO), was identical to the previous killings. And three: The suspected lead assassin, a female, appearing to be pregnant, was photographed by my surveillance team to be in the vicinity at the time of the assassination. Based on verified evidence, which was based on eye-witness accounts of pedestrians who were in the immediate area, along with comparative facial analysis by expert personnel from the office of the Central Intelligence Agency (CIA), the assumed-pregnant female is the lead assassin. In conclusion, the suspected assassins, who have appeared to be pregnant for ten to twelve months—well over a nine-month pregnant cycle—are not pregnant, but are wearing prosthetics to appear pregnant, which allows them to gain close contact with their targets, which explains the close-range kill shots to the victims," concluded Captain Gilmore. "I recommend that I be temporarily assigned to the Saigon Intelligence Division. Based upon history of MO, the assassins will be striking within the next two months."

General Drinkwater asked, "Captain Gilmore, do you have your sniper team with all of the required sniper equipment in place? If so, be prepared to depart for Saigon in the early morning hours of tomorrow," ordered General Drinkwater.

"Yes sir. I do. And yes sir. I will be prepared to depart tomorrow," John David said.

"Captain Gilmore, I accept your recommendations," the General said, as he concluded the meeting with a salute.

Immediately following his top secret briefing with General Drinkwater, Captain Gilmore, using a secure radio frequency, made a call to Hao Hong, and informed her of his assignment to Saigon, which was approximately ninety miles from her home, and that would allow them to visit with each other frequently.

An excited Hao Hong closed with a vow. "I love you, John David." She immediately informed her brothers of the good news, and that John David would be visiting more often, and they would be able to challenge him with their improved chess skills.

On March 9, 1966, at approximately 1600 hours, when the pedestrian rush hour traffic was at its peak, Captain Gilmore's assassin surveillance team caught sight of the suspected lead assassin, and immediately summoned Captain Gilmore, who was analyzing data in the quietness of his combined work and bedroom space of his kitchenette apartment, which was one floor down from the observatory and sniper's nest.

Having made secure radio voice connection with General Drinkwater, who would give the kill order, the following transpired.

Captain Gilmore said, "General Drinkwater, sir, I have a visual on the target. The distance is 250 yards. Wind is blowing from the north, at a speed of three to five miles per hour with no gusts. Temperature is 78 degrees Fahrenheit, and humidity is 34 percent. Elevation rise is 98 feet 3 inches. Medical staff, with OBGYN surgeon, is within two minutes away from the target, as a precaution for, if the assassin is pregnant, the unborn fetus can be saved with an emergency caesarian surgical procedure. I will execute, on your command, a kill shot to the head, eliminating danger to the fetus, if target is pregnant. Sir, permission to neutralize the target."

"Captain Gilmore, order to neutralize target, granted."

Captain Gilmore squeezed the trigger. "Sir, kill shot to the head, taken! Target is down! OBGYN medical team is enroute. Radio channel is off," concluded Captain Gilmore.

Later that evening, at approximately 2000 hours, the OBGYN medical team reported to General Drinkwater that the assassin was in-fact *not* pregnant, and was robed in artificial pregnant prosthetics, and a .22 caliber hand-gun was recovered at the scene. After careful testing and analysis of the recovered weapon, ballistic experts certified the weapon to be the same weapon used in five previous assassinations.

Later, during the war against the Viet Cong, the CIA was charged with the investigation of the female assassination team, who were neutralized. And the spy who was imbedded deep inside of the Saigon Intelligence Division was identified, court martialed, and found guilty of aiding and abetting the enemy. The mole/spy was sentenced to life, and was currently imprisoned at the military prison at Fort Leavenworth, Kansas.

With his sniper team successfully carrying out the methodical and arduous mission of neutralizing the female assassination team, who were responsible for an estimated eleven assassinations of American Intelligence officers, General Drinkwater issued orders for a three-day pass for Captain Gilmore and his sniper team.

Along with his three-day pass, General Drinkwater issued orders for the Special Forces team, which had accompanied Captain Gilmore on previous occasions, to escort Captain Gilmore to the Cu Chi Hamlet for the purpose to visit with Hao Hong.

Chapter Twenty-One

Til Death Do Us Part (TDDUP)

O n the early hour of 0600 hours of March 15, 1966, and the twelfth birthday of Dung, Hao Hong's youngest brother, Captain Gilmore, escorted by his personal Special Forces team on a forty-five-minute flight in a U-H1 helicopter, safely landed on the front yard property of Hao Hong and her brothers.

Having used the communication system to alert Hao Hong of his visit, Captain Gilmore, along with his escort team, was happily met by his adopted family.

While John David and his future wife were in a passionate embrace of hugs and kisses, he managed to wish Dung a happy birthday and presented each boy with unassembled gifts of model airplanes and automobiles, while not forgetting to wish Duc, the oldest brother, a belated happy thirteenth birthday.

In the meantime, Captain Gilmore's escort team was busy unloading four, ten-gallon coolers that were filled with ice, and six cases, containing twenty-four cans each, of Coca Cola soft drinks, which were a favorite of both Duc and Dung.

Alone together in Hao Hong's bedroom, John David positioned himself on to one knee, and while bowed, removed from the right-side pocket of the military fatigues that he was wearing, the one-half carat diamond with two, one-quarter, blue sapphire begets that were mounted on a gold-band, engagement wedding ring. The ring his mother, Rose, had worn, and had requested, before her untimely death, to be given to the first sibling to be married in holy matrimony. With this precious gift, and as memories of his dear mother flooded his mind, John David asked, "Hao Hong, will you marry me?"

"Yes, John David, I will marry you!"

John David slipped the ring onto Hao Hong's finger then placed his head on to the abdomen of his future wife, and noticed a small swelling in the belly of Hao Hong.

"I am pregnant," Hao Hong confessed, with tears streaming from her eyes, as she was nervously shaking.

"Why didn't you tell me?" replied John David, while still on one knee, with his forehead pressed against the abdomen, which was carrying his unborn child. John David rose from his kneeling position, and began to comfort his future wife while tasting Hao Hong's salty tears on his lips.

"There are thousands of Amerasian children in Vietnam, who were fathered by American soldiers, who left them with their mother to raise alone when they returned to the United States!" Hao Hong exclaimed with tears streaming down her cheeks. She admitted that she was afraid that John David would leave her alone to raise their child, along with the responsibility of raising her young brothers.

"Hao Hong, my darling, I am in love with you. I love you with all of my heart! God gave you to me. You are the first woman for me to make love with, and you will be the only woman whom I will make love with! Remember, darling, that I am the first man for you to give yourself to. I will never leave you and my baby that you are having for me," John David vowed. He summoned his future brothers-in-law, and announced the pregnancy of their sister, and that they would be getting married on April 9, 1966, on Hao Hong's 23rd birthday.

While sitting at the dining room table, and with each member of John David's adopted family enjoying a Coca Cola over ice, plans were made for Hao Hong and her brothers to travel, with the Special Forces escort, to Saigon during the week of Hao Hong's birthday, where they would be joined in holy matrimony in the US Embassy's chapel.

With a muffled giggle, Duc asked his sister, "Did you get pregnant while we were at the arcade, and at the movies during our Christmas vacation?'

"Or did it happen on the night when you went into Captain Gilmore's bedroom while wearing a long top coat?" noted Dung mischievously, as the brothers giggled, while challenging John David to a game of chess, and giving their permission for the husband and wife to-be to sleep in the same bed while John David was enjoying a three-day pass.

In quiet bashfulness, Hao Hong refused to acknowledge her brothers' curiosity as to when she became pregnant.

During the three days in the Cu Chi Hamlet, John David gained knowledge of the economic system that members of the hamlet had established. Each family produced and traded their products to sustain a livelihood. Hao Hong and her brothers raised chickens and ducks, along with the produce from their garden. All members were responsible for maintaining the health of the fish pond, along with the tilling, planting, and harvesting of the rice crops. All surplus from their production was sold for profit on the open market in Cu Chi City.

The home of each family was supplied with a limited amount of electricity to power a small refrigerator-freezer and a window fan.

With the day light hours filled with assisting Hao Hong's brothers with their chores of caring for the chickens and ducks, along with the gathering of produce from the garden, and harvesting fish from the hamlet's fish pond, and the occasional butchering of a chicken or duck for daily food, John David began planning for the construction of an additional room to the family's home, which would be needed for the soon-to-be addition of the new baby.

With permission from her brothers, Duc and Dung, for she and John David to sleep together, the now-engaged couple would wait for late night and early morning hours, hoping for her brothers to be asleep, before making love to renew their vows and commitment to each other.

On March 17th, at 1100 hours, after a warm embrace, and a happy but emotional goodbye to his adopted family, Captain Gilmore and his Special Forces Escort Team returned to Saigon, where on a secure radio call, Captain Gilmore informed General Drinkwater that Hao Hong was pregnant. He requested, permission to marry her, and for General Drinkwater to officiate the ceremony, which would be held in the chapel, at the US Embassy in Saigon.

"Permission granted," responded General Drinkwater, while noting with a humorous tone in his voice, "Captain Gilmore, when a handsome young military officer, and a beautiful young woman sleep with each other, they invariably will produce a baby!" He assured Captain Gilmore that he would count it as a blessing to officiate the wedding ceremony, and that he would order his staff to make all of the necessary arrangements for travel and hotel accommodations for Hao Hong and her brothers.

Escorted by the previously assigned Special Forces team, Hao Hong and her brothers, after making arrangement with the hamlet's chief to care for their chickens, ducks, and garden, arrived at their US Embassy's secure hotel, on the morning of April 7, 1966, two days before she and John David would be united in marriage.

The early check-in to their hotel allowed Hao Hong and her brothers to shop at the embassy's exchange store for an appropriate wedding dress.

At 1400 hours on April 9, 1966, which was a cool spring day with a high-blue sky and an outdoor temperature of 72 degrees Fahrenheit, along with comfortable humidity, bride, Hao Hong on her twenty-third birthday, along the side of her groom, Captain John David Gilmore, stood before Lieutenant General Thornton "Thorny" Drinkwater, with Duc and Dung as witnesses, to take their wedding vows of holy matrimony.

The wedding gown, which bride Hao Hong Nguyen had purchased two days earlier, and clearly revealed her fourth month of pregnancy, was a cream-colored, full-length gown, with long sleeves, and white, low-heeled shoes. The groom, Captain John David Gilmore, with a smile of joy defining his face, on this very special day, wore his Air Force dress blue uniform, with highly-shined black shoes. The bride and groom, following the instructions of a uniformed Lieutenant General Drinkwater, said, "I do," to the following vows of holy matrimony:

"Captain John David Gilmore, do you take Hao Hong Nguyen, to be your lawful wedded wife in health and in sickness, and in richness and in poverty, until death do you part?"

"I do."

"Hao Hong Nguyen, do you take Captain John David Gilmore, to be your lawful wedded husband in health and in sickness, and in richness and in poverty, until death do you part?"

"I do."

"Captain Gilmore, you may now place the wedding ring on to the left-hand ring finger of your bride, and Hao Hong, you may place the wedding ring on to the-left hand ring finger of your groom; these rings, which are a continuous and perfect circle, represent perfect and continuous commitment and fidelity between husband and wife, and with the authority vested in me, as a lieutenant general of the United States Air Force, I, Lieutenant General Thornton 'Thorny' Drinkwater, pronounce Captain John David Gilmore, and Hao Hong Nguyen, husband and wife. Captain Gilmore, you may now kiss your wife, Mrs. Hao Hong Gilmore!"

On returning to their hotel bridal suite, and before allowing her brothers to explore the grounds, various restaurants, and arcade on the campus of the US Embassy, the adopted family enjoyed a serving of the wedding cake, along with a can of Coca Cola, instead of a glass of Champaign, as Hao Hong, mother to be, was refraining from the consumption of alcohol.

Hao Hong instructed her brothers that the movie they were to attend would be ending at approximately 9:00 p.m., and they were to immediately return to their hotel room, which adjoined the wedding suite that she and their new brother-in-law were occupying.

At approximately 5:00 p.m., Duc and Dung departed from their room to attend a movie at the embassy's theater, which allowed Hao Hong and John David to consummate their marriage. Slowly undressing each other, the sexually aroused newlyweds embraced and consumed in passionate kisses, and while lying naked on the bed, blessed their marriage, as Hao Hong positioned herself onto her knees, in an effort to protect the unborn baby that was growing in her belly, and allowed John David's throbbing erection to enter into her wetness from the rear. Exhausted, drained, and relaxed, Hao Hong, while passionately kissing the open mouth of her husband, confessed, "I did not know that a pregnant woman could have so many orgasms!"

"And I did not know that a pregnant woman could be so wet, and horny! Baby, you know that your man, is not finished?"

"I hope not," responded Hao Hong, as she gently fondled her husband's throbbing erection, while on her knees, submitted to him in hopes that each of them could be satisfied before her brothers would return from attending the movies. Exhausted from their insatiable appetite for love making, the newlyweds fell into a relaxed, and deep sleep, and failed to hear the return of Duc and Dung to their room at approximately 2100 hours, as their sister had ordered, immediately following the ending of the movie they were attending.

On the following morning, a Sunday, and after having their breakfast served in their respective rooms, John David embraced his adopted family, and requested a promise from his brothers-in-law, that they would take care of their sister, his wife. With a bitter-sweet goodbye, John David watched his wife and adopted family board the U-H1 helicopter for their return to their home, in the Cu Chi Hamlet.

While preparing to return to the duty of war, Captain Gilmore pondered in his heart, with the stench of war engulfing him, and the responsibility of a wife, and her dependent young brothers, along with a child soon to be born, would he ever return to Wyatt, Missouri, his home town? With that thought in mind, John David began to subconsciously devise a plan to remain in Vietnam for the purpose to raise and take care of his family, a value which had been instilled into him by his father, Alonzo Gilmore.

The character and picture of the Vietnam War was fully developed during the early months of 1966. The death toll for American soldiers was estimated to be eighteen to twenty per day, seven days a week, which amounted to approximately seven thousand for the years of 1966-67. Those who were severely injured reached an astounding total of twenty-one thousand per year. Death, mayhem, and military failures were taken for granted, and accepted as the cost of war.

On the early morning of September 18, 1966, Captain Gilmore, while flying his F-104 fighter jet, in the support of B-52 bombers, was credited with his third shoot-down of a T-38 Trojan Fighter of the Vietnam People's Air Force (VPAF). There was no celebration. It was routine. It was accepted, as just another foul-smelling day of war, which had come to be a normal, and another odorous day, which engulfed South Vietnam. The sickening smell was caused by human waste, burn pits, which was fueled by garbage produced by humans and carcasses of dead animals that were collateral damages of the raging war. The plumes of smoke, which were carried by the prevailing winds to all corners of Vietnam, numbered into the hundreds, if not thousands.

The sickening smell reminded Captain Gilmore of the odor on the Smithfield Plantation during the summer of 1950. Hogs on the Smithfield Plantation, and throughout the Bootheel of Southeast Missouri, contracted the highly contagious cholera disease. To prevent an endemic spread of the disease to adjoining states, the animals, along with the barns, and hog pens, were euthanized, and destroyed in burn pits. The burning of hog flesh, along with the barns and pig pens, which the animals inhabited, lasted for approximately two months, and produced a sickening odor that permeated the memory of John David. This dreadful memory was awakened with the death and smell that the Vietnam War produced.

The timing of Captain Gilmore's third shoot-down of an enemy aircraft coincided with the birth of his first child. While her husband was on assignments, which included the neutralization of enemy snipers and fighter jet missions to support B-52 bombers, General Drinkwater assumed personal responsibility to ensure that Hao Hong would receive the necessary medical that was needed during her pregnancy and delivery

On September 21, 1966, after leaving the responsibility of her home, and brothers, under the supervision of the hamlet's chief, Hao Hong was transported to the maternity ward of the US Embassy's hospital in Saigon, where on September 23rd, she gave birth to a healthy, six-pound, baby girl. On the following day, Captain Gilmore arrived at the bed side of the new mother and embraced his wife and held his first-born child in is arms.

Before being discharged from the embassy's hospital, the proud parents had to decide on a name, and nationality for the purposes to secure a birth certificate, social security number, and record the child as a dependent of a US Military officer.

For nationality, the first-time parents were at a loss, and filled with laughter as they attempted to check the block defining, nationality. "I am half French, and half Vietnamese, and John David, you are White American, Negro, and Creole. So, what does that make our baby?"

The proud parents mused in laughter, as they decided to write in "American".

"John David, my dear, what name shall we give to our precious, baby girl?"

"We shall name her Rose! My mother's name was Rose."

"Rose was your mother's name? Wow! That is the name I chose for her, because when my name, Hao Hong, is translated to English, it means, *rose!*"

After three days of convalescing on the maternity ward, and filing the required documents to record birth, and US Military dependent, and beneficiary, Hao Hong was transported on a U-H1 helicopter, by Special Forces escort, to her home, along with her baby, Rose, where both were welcomed by proud uncles Duc and Dung.

With the birth of baby Rose, and the construction of an additional room to their shared home, General Drinkwater allowed for Captain Gilmore to remain on assignment at Saigon's Air Force Military Intelligence Division, where he would be allowed to visit with his wife and family on a frequent schedule.

On January 15, 1967, General Drinkwater informed Captain Gilmore that he had waived the four-year requirement of time and grade at the rank of captain before being eligible to be promoted to the rank of major of the United States Air Force, and that he could immediately begin to wear the gold oak leaf insignia, which represented the rank of major. In addition to being promoted to the rank of major, General Drinkwater also informed, now Major Gilmore, that he had been moved to the top of the list to attend the National Aeronautical Space Administration (NASA) training facility to become an astronaut.

During the days, weeks, and months of 1967, the Vietnam War grew more intense. The number of American casualties increased day by day.

With the fighting growing in its brutality, especially with the Viet Cong utilizing guerilla and sniper tactics, Major Gilmore was called on to neutralize three enemy sniper teams, as well as flying his regular fighter jet missions in support of the B-52 bombers, where he was credited with the shoot-down of an additional enemy fighter plane, bringing his shoot-down of enemy aircrafts to a total of four.

In addition to his sniper assignments, and fighter jet support missions, Major Gilmore volunteered to fly U-H1 helicopter medical evacuation missions. On January 28, 1968, at approximately 0100 hours, and in the darkness of a cold and rainy night, and at the height of the Tet Holiday, inspired offensive, the U-H1 medical evacuation helicopter, which Major Gilmore was solo piloting, had a mechanical failure, and crash-landed approximately twenty-five miles southwest of the 25th Helicopter Battalion, which was stationed at the US Military Army Base, at Cu Chi, Vietnam, which shared the name of the hamlet where Major Gilmore's wife, daughter, Rose, and brothers-in law lived.

Recognizing that this crash landing was by Divine intervention for him to fashion a scenario for him to be counted as missing in action (MIA), Major Gilmore chose not to radio a Mayday distress, which would have recorded coordinates and location of his crashed helicopter.

While suffering with the pain of several broken ribs, and a deep cut below his right knee, and before destroying the crashed helicopter with an explosive device, Major Gilmore made a secure radio call to his wife, Hao Hong. Instructing her that it was now the perfect opportunity to put into action the plans they had made for him to be counted as MIA so that he could remain in Vietnam, he gave to her the geographical coordinates of his location, along with compass readings from their home to his position, and that while injured, he would move himself in her direction to be rescued.

Hao Hong, understandably nervous and frightened, and three months into her second pregnancy, assembled a rescue team from her hamlet, which included her oldest brother, Duc, along with two females, and four males. Leaving her youngest brother, Dung, to care for baby Rose, the rescue team, utilizing two, three-wheeled, motorized vehicles, which were normally used to transport produce and animals to the local market for sale, set out to rescue Major Gilmore.

As she had prayed in faith that God would send to her a husband, Hao Hong, once again, in faith prayed that God would now keep her husband safe and deliver him back to her.

With both Hao Hong's rescue team, and Major Gilmore moving to meet them by using the compass reading, and on the same line to meet, God answered the prayer of a faithful wife, and Major Gilmore was rescued, just as the morning sun was rising. He was safely returned to his home and daughter in the Cu Chi Hamlet.

On January 31, 1968, at the local time of midnight in Wyatt, Missouri, Major Gilmore, using the military-grade radio that had been installed in his home, along with his technical skills, patched together a UHF signal to the telephone land line of his brother Jacob, and with a sworn vow of secrecy from Jacob not to share with anyone, including their sisters, shared what had occurred in his life during the past eighteen months, and based on the teaching of their father, Alonzo, to never desert family, that he had made the decision to remain in Vietnam, and that he had been officially reported missing in action.

"I pray to God, that I will again see you, Jacob, and our sisters, Martha, Sarah, and Rachel, and our pet mule, Blue, but for now, FAREWELL."

Epilogue

(WHERE ARE THEY NOW?)

On June 3, 1968, John David and Hao Hong welcomed fraternal twins into their family. Naming the new addition after her mother and father. The girl, who was born first by two minutes, was named Brigitte; the boy was named Binh. After the birth of the twins, Hao Hong had a surgical procedure to tie-off her fallopian tubes, to prevent giving birth to additional children; therefore, preventing any possible suspicion as to the validity of Major Gilmore's MIA status.

John David's beneficiaries continued to receive his allotment of payment for a major of the US Air Force. The financial compensation was used to upgrade the gas-powered generators that supplied electrical power to the Cu Chi Hamlet. John David also used his military compensation to construct a four-hundred-square-foot, dual purpose, worship center and school, where he and Hao Hong served as pastor and teacher.

On July 26, 2021, the date of his birth, John David peacefully died while asleep at the age of eighty. And at the time of this writing, Hao Hong, along with her oldest brother, Duc, and twins, Brigitte and Binh, continues with the pastoring, and teaching at the church-school that was founded and built by John David. The facility was named "Saint James," in memory of his uncle, James, who died after running away from the Gilmore Plantation of Tupelo, Mississippi with his brothers Alonzo and Benjamin.

Rose, the oldest child of John David and Hao Hong, lives in Hollywood, California with her aunt Sarah, and works as a professional model.

Dung, Hao Hong's youngest brother, became an officer in the Vietnamese Army.

Jacob Gilmore never revealed John David's secret. He married Leila Burnett and they were blessed with seven children, five boys, and two girls. As of this writing, Jacob and his family continue to live in Wyatt, Missouri, and are successful farmers.

Martha received her doctorate degree in education, and as a tenured professor, continues to lecture at HBCU colleges and universities.

Sarah became an A-list actress, and as of this writing, she occasionally does bit parts in movie and television productions.

Rachel became an OBGYN physician, and as of this writing, continues to practice medicine.

Blue, the pet mule, died in early fall of 1975. He was forty-four, and as requested by Rose Broussard Gilmore, was buried under the giant pecan tree, which he stood beneath for its shade.

Benjamin died on August 4, 1980, exactly one month after his 76th birthday. He was survived by Beatrice, his wife, and three children. He remained employed with the meat packing company that gave him his first job upon his arrival in Chicago, Illinois.

Ada, the mother of Benjamin, and Flora, the mother of James opened a small soul-food restaurant on the south side of Chicago, which is still managed by the children of Benjamin.

Ada and Flora both died in 1950. Not ever having their birth recorded, it is believed that, on their death, they were sixty-five.

Mrs. Adaline Smithfield died in her home town of New Orleans, Louisiana in the summer of 1960. She was eighty-five. Her remains were transported to southeast Missouri, where she was buried in the Mule Ridge Cemetery, alongside of her husband, Hunter Smithfield Sr., and son, Hunter (Little Hunter) Smithfield Jr.

Lieutenant General Thornton "Thorny" Drinkwater, retired from the United States Airforce one month after certifying John David Gilmore, Major, United States Air Force, to be missing in action.

Jonathan Gilmore's body is still buried in South Korea. Jacob, his brother, continues to petition the Office of the Veteran's Affairs to bring home the remains of his brother, to be buried in the Mule Ridge Cemetery, along the side of their father, Alonzo, and mother, Rose.

Charlotte Rose Broussard, as of this writing, has not been located.

www.ingramcontent.com/pod-product-compliance
Lightning Source LLC
Chambersburg PA
CBHW020237130626
46549CB00005B/1937